THE NICK OF TIME

THE NICK OF TIME

Politics, Evolution, and the Untimely

ELIZABETH GROSZ

Duke University Press Durham and London 2004

Printed in the United States
of America on acid-free paper ∞
Designed by Amy Ruth Buchanan
Typeset in Minion by Keystone
Typesetting, Inc. Library of Congress
Cataloging-in-Publication data appear
on the last printed page of this book.

CONTENTS

ACKNOWLEDGMENTS

This book would not have been possible without the help, support, and input of many institutions and individuals. I owe a debt of gratitude to three universities in particular. The earliest researches for this book were undertaken at the Centre for Cultural Studies and Critical Theory at Monash University in Melbourne, Australia. Without the support and encouragement of faculty and students there, on whom I rehearsed the earliest and most incoherent ideas developed here, I would have had no idea where to go next, what to concentrate on, what to eliminate, what I knew enough about already and what I knew I didn't yet know, what was then beyond me, what I had to acquire in order to make sense of what, for me, was a new and unfamiliar topic.

The first two sections of the book were written while I worked in the Departments of Comparative Literature and English at the State University of New York at Buffalo. I am particularly grateful to the graduate students in those departments, who suffered through a number of courses in which I developed and refined much of the material that would find itself, I hope in improved form, here. These improvements and refinements were the results of the feedback and queries of these smart and critical students. My stay at UB was particularly facilitated by the kind and generous support of Ms Theresa Monacelli and through the intellectual stimulus provided by my colleagues there, Professors Mimi Long, Jill Robbins, Isabel Marcus, Carol Zemel, and Joan Copjec. My thanks to them for their generosity, kindness, and support during the period of my tenure at UB.

The book's final sections were completed while I worked in the Women's and Gender Studies Department at Rutgers University. I would like to thank especially Ms Joanne Givand and Professors Joanna Regulska, Barbara Baillet, and Harriet Davidson, who, each in her own way, made my transition to Rutgers as smooth and seamless as I had dared hope. I am grateful to the department and to Rutgers University for providing me the stability and confirmation, not to mention leave, I needed to complete the book.

While institutions provide the structure of material and economic support that enable books to be written, it is more personal, sometimes indirect or oblique, relations with individuals that drive one to write a particular book in a particular way. I am extremely grateful to a small circle of friends, colleagues, and family who have supported me while I pursued the sometimes crazy and often inchoate paths that led me to produce this text, particularly Judith Allen, Sue Best, Pheng Cheah, Sally Munt, Tony Nunziata, John Rajchman, Jacqueline Reid, Gai Stern, Jen St. Clair, Mary Gross, Tom Gross, Irit Rosen, Tahli Fisher, Daniel Gross, and Mia Gross. Claire Colebrook deserves special mention for her incisive, detailed comments and helpful suggestions. The book has definitely improved as a result of her suggestions. My thanks, as always, to my mother, Eva Gross, for being there, and for understanding that sometimes it is all right to not understand what her daughter has been up to. I want to acknowledge the particular place of Nicole Fermon in the production of this book. Without her support and belief, there would be no point in writing, especially writing about the need to affirm the new, the future, the unknown. She made the future seem an exciting place to want to be, to see, and to help produce. This book is dedicated to the memory of my father, Imre Gross, who made me understand, more than anyone, how precious time is.

ABBREVIATIONS

Charles Darwin

DM	*The Descent of Man*
OS	*The Origin of Species*

Friedrich Nietzsche

BGE	*Beyond Good and Evil*
EH	*Ecce Homo*
GM	*On the Genealogy of Morals*
GS	*The Gay Science*
"OUL"	"On the Utility and Liability of History for Life"
TI	*Twilight of the Idols*
WP	*The Will to Power*
Z	*Thus Spoke Zarathustra*

Henri Bergson

CE	*Creative Evolution*
CM	*Creative Mind*
ME	*Mind-Energy*
MM	*Matter and Memory*

INTRODUCTION: TO THE UNTIMELY

This book is an exploration of how the biological prefigures and makes possible the various permutations of life that constitute natural, social, and cultural existence. Its goals are interdisciplinary: to explore a series of questions about the movement of time that overlap with and are the shared concern of a number of different disciplines: philosophy, politics, history, the social and natural sciences, cultural studies, feminist, antiracist, and queer politics, the visual and plastic arts. It focuses on the space between the natural and the cultural, the space in which the biological blurs into and induces the cultural through its own self-variation, in which the biological leads into and is in turn opened up by the transformations the cultural enacts or requires. It explores the region between biology and culture; between bodies and sociotemporal organization, between the sciences of life and the study of social organization—a philosophical exploration on the cusp of science studies on the one hand and political theory on the other.

Biological organization, whose morphological structures engender the variety of life in all its forms, instead of ensuring that life conforms to existing social categories, boundaries, and limits, instead of containing existence to what is or has been, opens up and enables cultural, political, economic, and artistic variation. Biology is a system of (physical, chemical, organic) differences that engenders historical, social, cultural, and sexual differences. Biology does not limit social, political, and personal life: it not only makes them possible, it ensures that they endlessly transform themselves and thus stimulate biology into further self-transformation. The natural world prefigures, contains, and opens up social and cultural existence to endless becoming; in turn, cultural transformation provides further impetus for biological becoming.

This book functions primarily as a reminder to social, political, and cultural theorists, particularly those interested in feminism, antiracism, and questions of the politics of globalization, that they have forgotten a crucial dimension of research, if not necessary to, then certainly useful for more incisively reformulating the concepts on which they so heavily, if implicitly, rely. It is written as a remembrance of what we have forgotten—not just the body, but that which makes it possible and which limits its actions: the precarious, accidental, contingent, expedient, striving, dynamic status of life in a messy, complicated, resistant, brute world of materiality, a world regulated by the exigencies, the forces, of space and time. We have forgotten the nature, the ontology, of the body, the conditions under which bodies are enculturated, psychologized, given identity, historical location, and agency.

We have forgotten where we come from. This is a double forgetting: of the elements through which all living things are born and live, a cosmological element; and of the specific body, indeed a chain of bodies, from which we come, a genealogical or maternal element. Life is this double debt, and its forgetting is perhaps the condition under which the living come to know the world though not understand themselves. The exploration of life—traditionally the purview of the biological sciences—is a fundamental feminist political concern, not because feminists must continue their ongoing suspicions regarding various forms of (male-dominated) biological research, but because feminists, and all theorists interested in the relations between subjectivity, politics, and culture, need to have a more nuanced, intricate account of the body's immersion and participation in the world if they are to develop political strategies to transform the existing social regulation of bodies, that is, to change existing forms of biopower, of domination and exploitation. We need to understand not only how culture inscribes bodies—a preoccupation of much social and cultural theory in the past decade or more—but, more urgently, what these bodies are such that inscription is possible, what it is in the *nature* of bodies, in biological evolution, that opens them up to cultural transcription, social immersion, and production, that is, to political, cultural, and conceptual evolution. We need to understand, with perhaps more urgency than in the past, the ways our biologies work with, and are amenable to, the kinds of cultural variation that concern politics and political struggle.[1]

We need to return to, or perhaps to invent anew, the concepts of nature, matter, and life, the most elementary concerns of the cosmological and the ontological, if we want to develop alternative models to those inscriptive and constructivist discourses that currently dominate the humanities and social sciences, in which the transformation of representation is the only serious political issue, and where the body is of interest only in its reflection through discourse, its constitution in representation, or its mediation by images. We need an account of what the social and the political, the individual and the sexual must use in the "construction" of identity, the body, or culture, its natural resources—if construction and inscription still remain relevant metaphors, that is, if this image of subject-constitution or -construction is to have any complexity or explanatory power. If the body is to be placed at the center of political theory and struggle, then we need to rethink the terms in which the body is understood. We need to understand its open-ended connections with space and time, its place in dynamic natural and cultural systems, and its mutating, self-changing relations within natural and social networks. In short, we need to understand the body, not as an organism or entity in itself, but as a system, or series of open-ended systems, functioning within other huge systems it cannot control, through which it can access and acquire its abilities and capacities. Our continuing studies of subjectivity and the body in the humanities and social sciences inevitably, if we go deeply enough, bring us back to the more complex and unsolved questions of the natural sciences, questions we had perhaps hoped to foreclose, sidestep, or ignore, but that now press upon our most intimate and subjective experiences with more and more urgency. It is precisely because we are interested in the social, the cultural, the historical, and the political forces shaping subjectivity or identity that we need to turn again, with careful discernment, to those discourses, once rejected by feminists and political activists, that place the body in the larger cosmological and biological orders in which it always finds itself.

What I explore in the pages that follow are philosophical models that underlie and can be extracted from much biological, particularly evolutionary, research. I concentrate on a series of key texts where biology and culture remained inextricably entwined—in the writings of Charles Darwin, Friedrich Nietzsche, and Henri Bergson—particularly on their insights about ontology, that is, about the reality of matter, space, and time, drawing from them some broad implications for how we might reconfigure political and cultural theory and struggle. In an earlier book, I wrote about the social inscription of bodies, the ways bodies are produced as cultural objects and as experiencing subjects through the active rewriting, and thus cultural unhinging, of biological materials.[2] What I did not adequately realize then, and what this book hopes to correct, is that without some reconfigured concept of the biological body, models of subject-inscription, production, or constitution lack material force; paradoxically, they lack corporeality. They have been stripped of their bodiliness and thus of their force, energy, and activity. I failed there to adequately address how living matter, corporeality, allows itself cultural location, gives itself up to cultural inscription, provides a "surface" for cultural writing—that is, how the biological induces the cultural rather than inhibits it, how biological complexity impels the complications and variability of culture itself.[3]

The body is not the object of investigation here; rather, my object is a more primitive, basic, and elusive concept that requires careful discussion, for without it, there can be no conception of what kind of corporeality a living body, let alone a human body, a sexed, raced, and historical body, is and could become. The object of preliminary investigation in this book—and I stress that this is very much an initial exploration—is time: its modalities, its forms, its effects on both inorganic and organic materiality. Bodies, living and nonliving, are spatial and temporal beings. I have focused in other works on the spatial dimensions of the subjective and the subjective dimensions of space.[4] This text is concerned primarily with the question of the ontology of time, duration, or becoming, the ontological implications for living beings of their immersion in the always forward movement of time.

The elaboration of a philosophical theory of time is no easy matter. Time is perhaps the most enigmatic, the most paradoxical, elusive, and "unreal" of any form of material existence. This may be why physicists, whose goal it is to explain material existence and render it predictable, have often relegated the perception of passing and change, the perceptible effects of time, to the status of subjective illusion, mere appearance, matter encoded by the attributes of human perception (see, e.g., Grünbaum 1973). This may also explain why those in the social sciences, whose very element is temporal, nevertheless aim for the measurable and the spatial in their researches, and why disciplines in the humanities, while continually evoking and attesting to the historical, tend to conceptualize it through the transformation of objects or subjects, considering the past through its capture in memory or in reading practices, and linking the future to the phantasmic, the imaginary, or the fictional. Nevertheless, time is not merely the attribute of a subject, imposed by us on the world: it is a condition of what is living, of matter, of the real, of the universe itself. It is what the universe imposes on us rather than we on it; it is what we find ourselves immersed in, given, as impinging and as enabling as our spatiality. We will not be able to understand its experiential nature unless we link subjectivity and the body more directly to temporal immersion, to the coexistence of life with other forms of life, and of life with things, that is, until we consider time as an ontological element.

Time is neither fully "present," a thing in itself, nor is it a pure abstraction, a metaphysical assumption that can be ignored in everyday practice. It cannot be viewed directly, nor can it be eliminated from pragmatic consideration. It is a kind of evanescence that appears only at those moments when our expectations are (positively or negatively) surprised. We can think it only when we are jarred out of our immersion in its continuity, when something untimely disrupts our expectations.[5] It is almost as if it overwhelms us with its pervasive force to such an extent that we cannot bear to think it, we prefer that it evaporates into what we can comprehend or more directly control. We "naturally" think of time through the temporality of objects, through the temporality of space and matter, rather than in itself or on its own terms. This is why it cannot be present or present itself, why we cannot look at it directly, why it disappears the more we try to grasp its characteristics. We can think it only in passing moments, through ruptures, nicks, cuts, in instances of dislocation, though it contains no moments or ruptures and has no being or presence, functioning only as continuous becoming. It is to the exploration of those nicks, disruptions or upheavals—events that disrupt our immersion in and provoke our conceptualization of temporal continuity, events that also make up the unpredictable emergences of our material universe—that this book is directed.

Time has preoccupied philosophical thought from the latter's inception. In a sense, it is the most pressing, the most philosophical of problems that living beings must encounter, yet the one least able to be mastered and controlled by life or understood through reason. Instead of containing and controlling time, life succumbs to its rhythms, direction, and forces, to the ever pressing forces of development, growth, and decay. Time inhabits all living beings, is an internal, indeed constitutive, feature of life itself, yet it is also what places living beings in relations of simultaneity and succession with each other insofar as they are all participants in a single temporality, in a single relentless movement forward. All forms of life must organize their receptivity to and their actions in an environment according to a temporal economy, whether they do so consciously or not. Philosophy, like other forms of knowledge, directs its focus to the movement of time, to duration, only extremely rarely and often under the influence of either scientific-mathematical or phenomenological-experiential models. Along with the sciences, philosophy tends to submerge time in representations of matter and space, to spatialize and visualize temporal movement in terms of the transformation of objects, when duration as such is capable of being experienced only directly, in its own temporal dynamics. It has eluded thought while nonetheless persistently intruding on it.

Although we *live* time continuously, although we are immersed in its movement so imperceptibly and naturally that our temporality, our irreducible movement forward, aging, is often unrecognized or automatic, we find it almost impossible to *think* or conceptualize temporal movement, to theorize it in its full implications. The ancient Greek philosophers confronted the question of time by linking its movement to mortality, change, and decay. As a remedy to its relentless and uncontrollable movement, they sought the immortal and the unchanging, the timeless, attainable through reason, through a distancing from the body, and, ultimately, through death. Yet within Greek thought, the fascination with transformation, movement, and change yielded dissenting positions, which reveled in flux and change rather than sought refuge from them, which understood time and change as real rather than illusory, and focused on the particular and the specific rather than the universal and the ideal. Heraclitus, Epicurus, and even, in their own ways, indirectly, Parmenides and Zeno, may be regarded as intellectual progenitors of the philosophical tradition I explore here.

Whereas philosophy has, since its inception, struggled with the question of time and how best to conceptualize it, this study focuses on three key figures in mid-nineteenth- and early-twentieth-century Western thought who revolutionized the ways we in the twenty-first century come to understand bodies in their fundamental intrication with time: Charles Darwin, the founder of contemporary biology; Friedrich Nietzsche, arguably the most notorious and misunderstood of all philosophers; and Henri Bergson, a philosopher whose fate has swung from the highest to the lowest and who today remains largely unread. Darwin, Nietzsche, and Bergson each provide us with raw materials, crystals of insight, regarding the unhinging of identity and the problematization of knowledges that acknowledging the full force of temporality implies. Although they do not present a single cohesive and integrated understanding of the intrications of time and life—there are many tensions, inconsistencies, and divergences in their respective claims—nevertheless, they do represent a particular orientation that brings philosophical thought side by side with scientific discovery, and enables them to interact with and affect each other without one functioning as the object of investigation of the other or as the metatheoretical judge of the other's relevance. This book does not aim to provide a philosophy *of* science, a philosophy that reflects on and adjudicates the value of scientific discoveries, or even a philosophy whose job is to justify and explain the sciences while leaving them autonomous and intact, but rather to provide some ingredients for a philosophy of life—perhaps even a philosophy of nature, though this concept is overburdened by the weight of nineteenth-century romanticism—which has its place alongside the empirical discoveries of the natural sciences and the affective and perceptual explorations of the arts.

There are, of course, many figures in the history of Western philosophy whose writings could be relevant to reconsidering the life of time and the time of life; each generation and philosophical tradition has wayward figures who insist on the reality of time and change. William James, Alfred North Whitehead, and the pragmatists on the one hand, and Hegel, Husserl, Heidegger, and the phenomenological tradition on the other, each in their various ways, insist on the reality of time, its functioning as a force with its own sometimes highly disputable characteristics.[6] No doubt each has something to contribute to a project such as this one. But here I concentrate on the strange mixture produced by bringing together Darwin, Nietzsche, and Bergson through a perhaps jarring encounter, that is, through linking each to the other by means of their allied concepts: life, evolution, and becoming.

For Darwin, life is essentially linked to the movement of time. He transformed the concept of life, in quite dramatic but commonly unrecognized ways, from a static quality into a dynamic process. In his writings, being is transformed into becoming, essence into existence, and the past and the present are rendered provisional in light of the force of the future. Life is construed as a confrontation with the accidental as well as the expected, a consequence of the random as well as the predictable. It is the response, the very openness, of material organization to the dynamism of time. In short, life is now understood, perhaps for the first time in the sciences, as fundamental becoming, becoming in every detail. Darwin makes it clear, indeed a founding presupposition, that time, along with life itself, always moves forward, generates more rather than less complexity, produces divergences rather than convergences, variations rather than resemblances. Descent, the continuity of life through time, is not the transmission of invariable or clearly defined characteristics over regular, measurable periods of time (as various essentialisms imply), but the generation of endless variation, endless openness to the accidental, the random, the unexpected. He thus makes temporality an irreducible element of the encounter between individual variation and natural selection, the two principles that, in interaction, produce all of life's organic and cultural achievements.

Darwin creates a real committed to a concept of temporal becoming. He creates a science in which history, and thus the eruption of unexpected events, is central, in which life is focused on and a response to local and global events, which are always unique and unrepeatable, which defy precise causal explanation, and which can provide explanation only at a certain level of generality—not precise prediction, which can calculate all the causal links constituting any event, but only the articulation of broad tendencies, which explain no individual in particular and understand species in terms of trajectories emergent from individual transformations. The movement of evolution is in principle unpredictable, which is to say that it is historical: related species in the past prefigure and provide the raw material for present and future species but in no way contain or limit them. The sciences that study evolution—evolutionary biology and genetics, for example—become irremediably linked, in spite of their current aspirations, to the unpredictable, the nondeterministic, the movement of virtuality rather than the predictable regularity that other sciences, such as physics and chemistry, tend to seek. The present and future diverge from the past: the past is not the causal element of which the present and future are given effects but the ground from which divergence and difference erupt.

Like all modes of historical interpretation or analysis, Darwinian evolutionary theory is fundamentally retrospective or reconstructive: given what exists now, we may be able to provide conjectured links and tracks that describe an evolutionary or temporal path to the present. But at a given moment in this history, it is impossible to predict what will follow, what will befall a particular trend or direction, let alone a particular individual, what will emerge from a particular encounter, how natural selection will effect individual variation, until it has occurred, until it is completed or provisionally arrested.

In recognizing the surprising, unpredictable, and mobile force of time on the emergence and development of the multitude of forms of life, Darwin brings the concept of the *event* to the sciences. Events are ruptures, nicks, which flow from causal connections in the past but which, in their unique combinations and consequences, generate unpredictability and effect sometimes subtle but wide-ranging, unforeseeable transformations in the present and future. Events erupt onto the systems which aim to contain them, inciting change, upheaval, and asystematicity into their order. Embedded in and incited by the force of unpredictable events, life evolves and transforms itself and, to some extent, its world, as a provisional mode of dealing with, responding to, the events that impinge on and affect it.

Darwin introduces indeterminacy into the Newtonian universe, a closed system in which matter is governed by a relatively small number of invariable, predictive laws (Newton 1999). If one could somehow take a snapshot of the Newtonian universe at any one moment, one could predict the future of any element within it, as well as its place within the detailed configuration as a whole. Newton posited a regular, predictable universe in which life, if it could understand and utilize those consistencies, would find itself at home, would tend to mastery, to understanding and control. Darwin sought to model his own scientific endeavors on such an enlightened and rational understanding of the role of science in rendering life safe, but what he produced was a very different account: life can be life only because the universe, at least as far as the living are concerned, is where it is never fully at home, where it can never remain stable, where it must undergo change both in itself, at the level of individuals, as well as over generations, at the level of species or populations. Operating at a different, a faster or slower rate of speed than much of the material universe,[7] life is always challenged to overcome itself, to invent new methods, regions, tactics, and goals, to differ from itself, to continually invent solutions to the problems of survival its universe poses to it, using the resources the universe offers it, for its own self-overcoming.

This book's first section is devoted to Darwin's own writing and to explaining his contributions to understanding the time of life. Chapter 1 explores Darwin's positioning of difference at the heart of his notion of species and the linkages he draws on and elaborates between the study of life and the study of languages and of economies, themselves vast, perhaps even defining, cultural systems of differentiation. Chapter 2 examines Darwin's postulate of the interactions of individual variation and natural selection as developed in *The Origin of Species* and *The Descent of Man*, and how they explain both biological and cultural emergence. Much of this material will be very familiar to those working in the biological sciences, yet it is important to introduce to readers from the humanities the intricacies and detail of Darwin's own writings, which, though popularized, are rarely read or referred to. Those already familiar with Darwin's writings may choose to skip these two chapters. Chapter 3 focuses on sexual selection and its role in the creation of sexual and racial morphological differences, drawing out Darwin's relevance to feminist and antiracist discourses. Between them, these three chapters introduce the reader to evolutionary theory and to the force of time on the emergence and elaboration of life into the various sexually bifurcated cultural and social forms that Darwin's model of evolution proposes. They demonstrate that Darwin remains, in spite of feminists' resistance to his work, one of the few thinkers of the nineteenth century to prefigure, not simply an egalitarian feminism, but a feminism, or perhaps more broadly, an ontology of sexual difference (see, e.g., Sayers 1982). He proposed that sexual bifurcation has become, through the unpredictable vicissitudes of biological history, the most successful and the least variable element in the descent of life: from the earliest elaborations of life's emergence on land, sexual difference is the strategy life has developed to ensure its maximum variation and proliferation. It is the very motor of life's self-variation, life's most ingenious invention for its own variability, regeneration, self-surpassing, and elaboration.

Although his relation to Darwin remains complex and highly ambivalent, Nietzsche is perhaps more Darwinian than he wants to admit. He, too, places unpredictable forces, temporal forces bound up with the future, at the center of his philosophy, which, like Darwin, he links to the functioning and affirmation of the inventiveness of life, life conceived without man as its culmination or center, life that unfolds through man rather than directs itself to man. Life is a form of self-overcoming, a form of affirmation, an excess or superabundance of opposing forces, whose internal will, what Nietzsche describes as the will to power, interprets and thereby transforms itself and its world. Rather than posing as a careful empirical observer whose role is to note down the details of animal existence, as Darwin the scientist did, Nietzsche renders Darwin's work truly philosophical. He returns Darwinism to the history of philosophy, which has resolutely repressed temporal flow, forcing it to confront physics and cosmology in ways Darwin himself resisted, and he draws out and makes explicit the moral and physical implications of a world without divine plan, a world of emergent order. He ontologizes and moralizes Darwinism; he makes his own version of Darwinism the beginnings of a philosophy of becoming. In turn, he produces his own Nietzschean version of "science," a kind of physics of forces, as the practical expression of his ontology. His theory of the eternal return is his response to Darwin's biological provocation to philosophy, the "gift" Nietzsche gives from his own excessive philosophy to the prevailing forms of knowledge, including dominant models of thermodynamics and social Darwinist biology, regarding the cosmos. What Nietzsche introduces to the Darwinian philosophical frame is not only that which concerned Darwin, that is, the well-adapted, what fits with its time, the anonymous or average member of a collectivity—the statistical or abstract entity, the organism, the typical individual so beloved by the sciences—but that which cannot be reduced to the average, the typical, or expected: the unique individual, the one who stands out, who is *exapted*, that is, adapted not so much to present circumstances as to the future. What Nietzsche focuses on is the out-of-place, that which marks itself as beyond or outside, the unaverage, the overman, the *untimely*.[8] The untimely is that which is strong enough, active enough, to withstand the drive of the present to similarity, resemblance, or recognition, for the untimely brings with it the difference that portends the future. The overman is the one who welcomes difference, the future, and its rewriting and transformation of the present, the one who is strong enough to seek his own erasure as man. This prophesied erasure of man must be of some direct relevance, however speculative, to feminist and other related political and theoretical struggles. Is the overcoming of mankind also the becoming of woman? To the overman, can one counterpose the overwoman? Is the overman also the overcoming of multiplicity, of those others that normative "mankind" has excluded? Or is the overcoming of mankind the generation of maximum difference?

Nietzsche elaborates a small space of excess that functions outside of natural selection, where life does not simply fulfill itself in surviving in its given milieu successfully enough to reproduce, but where it actively seeks to transform itself, where it refuses reproduction and instead seeks transformation. Life is still bound to the forward movement of temporality, but this movement is not that of conformity to one's nature or position, but welcomes what transforms and adds to one's nature and position. Life is not equally expressed in all its varied forms, as Darwin believed, but is most expressed in a life that makes activity its goal, that seeks to overcome itself and its present by welcoming whatever the future may hold. Nietzsche makes the force of time the very method for the evaluation and transformation of life.

Nietzsche not only transformed prevailing philosophical conceptions of subjectivity, identity, and life; he transformed philosophy itself. Philosophy as a mode of reflection on concepts was recreated by Nietzsche as a philosophy that actively tests and evaluates, that functions as a participant in power relations, not only observing the force and power of the objects it investigates, but adding its own force and power to them. Philosophy, and all forms of knowledge, including the natural sciences, are not just modes of observation, the impartial recording and analysis of information; they are active modes of interpretation, active forces of transformation, more or less dynamic interventions into what is being observed. Knowledge is now at play in the world, not a respite from it. The second section of the book is devoted to outlining Nietzsche's complex relation to evolutionary thought and developing his alternative—the eternal return—to Darwin's conception of natural selection. Chapter 4 outlines and renders explicit Nietzsche's critique of Darwinian evolution; chapter 5 develops his account of the will to power as a cosmological and moral principle; and chapter 6 explores his concept of the eternal return and the overman as his response to the Darwinian account of the evolution of man: as man has evolved from lowly animals, so too shall man himself be superseded, overcome by the overman.

There are many resonances between the writings of Bergson and those of Nietzsche on the one hand and Darwin on the other. Bergson, too, develops an untimely understanding of the task of philosophy. For him, philosophy is no handmaiden to the sciences, nor is it tied to the imaginative projects of literature and the arts. Rather, it is a practice, functioning alongside these other cultural practices, drawing from them, which adds to these practices what they must leave out. It is the practice of conceptualizing the immersion of materiality, whether organic or inorganic, in duration. Like Nietzsche, Bergson reconceptualizes, and problematizes, the place of philosophy in the order of knowledges; like Darwin, he reconceptualizes, and complicates, our understanding of life in its open-ended and unpredictable becoming. He focuses on the ontological dimensions, the metaphysical implications, of life, the ways life resists science to reveal itself in its concrete specificity only through an arational intuition whose logic or economy is linked to the very self-proliferation of temporal movement. For Bergson, as for Nietzsche, philosophy is an activity of incitement to action, as well as a space for the resolution of that which underlies action and makes it larger than itself. It is because we act in a world whose complexity we cannot fully perceive or comprehend, which we cannot fully anticipate or control, that we invent, we make more of ourselves than life makes of us. This inventiveness is not the productive effort of intellect, which is bound up with control of what exists in the present, that is, with the actual, but is a leap outside ourselves that emerges or has value only in the future. It is a gamble, like Nietzsche's throw of the dice, or an accident, like Darwin's natural selection.

Bergson remains closer to Darwin than to Nietzsche, though his interest is not primarily scientific, but philosophical, that is to say, metaphysical. Yet he is concerned, in ways that Darwin and Nietzsche are not, with the natural articulations of life's evolutionary movement, the particular, historically specific ways in which life divides and complexifies itself according to the different ways that living beings experience the events that provoke and stimulate their existence and profusion. Plants, insects, and vertebrates are life's most prominent morphological modes of response to the different events, problems, and provocations that nature addresses to the living. Instinct, intellect, and intuition are the resources life develops from its own chemical and morphological particularities to deal with, to invent, heritable or collective responses to these provocations. Like Nietzsche, Bergson proposes a new way of understanding philosophy as the supplement of that which other knowledges cannot know: it is the form of knowledge that affirms the interconnectedness of being in its temporal embeddedness; it is the knowledge that invents and sustains intuition. It has an affinity with literature and the arts no less than with the sciences, but it remains its own particular form of knowing, focused on the elaboration of concepts that elucidate as best they can the ongoing continuity of time.

Bergson's rewriting of evolutionary theory occupies the third and largest section of the book, again in three chapters. Chapter 7 introduces Bergson's conception of duration as qualitative difference in kind and explores the qualitative differences between the past and the present. Chapter 8 details his conception of creative evolution, his philosophical elaboration of the implications of Darwinism. And chapter 9 presents his understanding of the evolutionary emergence of instinct, intelligence, and intuition, bringing with it the question of the opposition between knowledge and life, between intelligence and intuition. As with Darwin and Nietzsche, Bergson too has an unrecognized relevance to feminists and cultural theorists: his notion of the virtual, embedded in his understanding of duration, is a crucial if commonly unrecognized concept for reconceptualizing the dynamics of political change and social and cultural upheaval. This is a concept that will develop increasing relevance as feminist theorists come to consider how change is developed and the new brought into existence.[9]

Darwin, Nietzsche, and Bergson make strange and discomfiting companions. Their interactions are most directly facilitated by two contemporary philosophers whose work is not the focus of discussion here, but whose contributions to rethinking the notion of becoming are without equal. Gilles Deleuze is perhaps one of the key figures to revive a nondogmatic interest in the primary texts of Nietzsche and Bergson, though his notoriously difficult writings do not provide readily assimilable interpretations of their respective contributions. His writings enable Nietzsche and Bergson to be placed in an exacerbating relation to each other, and for both to force a kind of confrontation with their mutual predecessor, Darwin.[10] He seeks what is untimely, what can be viewed only through a nick or crack, in the writings of each. Luce Irigaray is the other philosopher whose writings undergird the study of the untimely developed here. It is Irigaray who insists on the irreducibility of sexual difference, a claim that finds startling confirmation in the writings of Darwin. She makes it clear that sexual difference is the condition of all life, not just in the present, but in the future to come. It is she who reminds Nietzsche of the watery maternal element his projects for self-overcoming entail and which they need to forget. And it is she who develops, perhaps more than any other writer since Bergson, a practical metaphysics that rigorously develops Bergsonian intuition as its primary method. Though their work deserves more detail than I can give it in this volume, I have attempted to run Deleuze's and Irigaray's concepts underneath my reading of the others, as a kind of (ghostly) guide to the untimely, the crack, the unexpected resonance, that each prefigures.

Through a careful reading of key texts of Darwin, Nietzsche, and Bergson, I believe that some of the most interesting and pressing questions in cultural, political, and artistic production can be addressed in ways previously precluded. Feminist, antiracist, and postcolonial discourses and struggles are not only directed to the attainment of specific short- and long-term goals; they, too, are as immersed in temporality as they are in corporeality. They are immersed in duration because bodily existence is endurance, the prolongation of the present into the future. Political and cultural struggles are all, in some sense, directed to bringing into existence futures that dislocate themselves from the dominant tendencies and forces of the present. They are all about making the future different from the past and present, in rupturing the continuity of processes through the upheaval posed by events. They are about inducing the untimely. The more clearly we understand our temporal location as beings who straddle the past and the future without the security of a stable and abiding present, the more mobile our possibilities are, and the more transformation becomes conceivable. The more we affirm the value of the nick, the cut, or rupture, the more we revel in the untimely and the more we make ourselves untimely. Political activism has addressed itself primarily to a reconfiguring of the past and a form of justice in the present that redresses or rectifies the harms of the past. It needs to be augmented with those dreams of the future that make its projects endless, unattainable, ongoing experiments rather than solutions.

PART I. DARWIN AND EVOLUTION

1. DARWINIAN MATTERS: LIFE, FORCE, AND CHANGE

Darwin's great novelty, perhaps, was that of inaugurating the thought of individual difference. The leitmotiv of *The Origin of Species* is: we do not know what individual difference is capable of! We do not know how far it can go, assuming we add to it natural selection.—Gilles Deleuze, *Difference and Repetition*

Ontology seems to be the forgotten or elided element of contemporary philosophy. The devastating critique of metaphysics that revitalized the natural sciences, helped generate the social sciences, and effectively transformed philosophy during the earliest years of the twentieth century has perhaps succeeded only too well in adjudicating not only the appropriate and inappropriate questions to which knowledge must direct itself, but in dismissing many questions that, it seems, we cannot do without, that we cannot *but* ask. Some of the most basic questions of ontology—What is matter? What is life? How do they link together? What are their relations of intrication?—need to be readdressed, perhaps not in the same terms in which they were originally considered, but in more contemporary language, which considers the social, historical, and political context in which metaphysics is invariably, if unconsciously, embedded. In the desire to abandon metaphysical presuppositions and to replace the apparently unanswerable questions of ontology with the more modest propositions of epistemology, self-consciously moving from the unknown to the knowable, shifting the ground from what exists to what we know, the inevitable ontological investments of discourses, the presuppositions they must make about the modes, types, and forms of existence they analyze, have remained unexamined, though the production of ontologies continues unabated. In distancing ontology ever further from epistemology, we lose the capacity to provide political critiques of epistemologies, for we lose access to what is outside, to the outside of knowledges, to what they leave out, transform, or cannot know. The more we focus on the production of knowledges, the less we are able to address the real, what is outside, what constitutes the gaps or flaws of existing knowledges.

I examine three interrelated philosophical clusters of concepts that are embedded in these central questions of ontology. First, I explore those ontologies in which matter is imbued with a dynamism or activity, in which nature is construed as force, provocation, activity, or incitement, rather than, as is the current fashion in feminist and cultural studies, where nature is considered an inert passivity onto which life, culture, and the human impose themselves.[1] Second, I examine those epistemological frameworks that actively affirm the perspectival orientation of all forms of knowledge, the historical, social, and sexual specificity of ways of seeing the world, ways of understanding the real, which accept that there may be a number of competing and possibly incommensurable epistemic models adequate to the richness of the real. Third, I examine those discourses that imbue time with an existence autonomous from space, from objects and from models that privilege causal or deterministic prediction, which see time in terms of the precedence of the future, time as an active, forward-moving force, a positivity that both coheres and transforms, that makes as much as it unmakes or decays. These three clusters of concepts—those linking matter, time, and (sexual and other forms of) difference—are embedded in their own ontological presuppositions, and between them, they may help us both to refigure and transform the stasis associated with most conceptualizations of ontology and to understand matter and life as dynamic forces, bound in various forms of cohesion, as modalities of difference. This conception of life as the mobilization of maximal difference links these abstract metaphysical questions to the concerns of contemporary politics, that is, to the productive destabilization of present social and cultural arrangements. Rethinking time and matter may help transform how we understand politics and political struggle.

This highly selective discussion of the intrication of life with time and matter begins with the key writings of Charles Darwin, who not only developed the theory of natural selection into a scientific research paradigm of unparalleled fruitfulness and success for nearly a century and a half, but who also produced a philosophical framework whose resonances have still not been properly understood, even today. There has been a great deal of attention devoted to Darwinism, to scientific developments and elaborations within biology and its cognate disciplines since the writings of Darwin himself, and a vast amount of published material has appeared under the rubric of scientific or empirical Darwinism. Darwinism has also had a powerful effect on literature, on cultural and artistic representations, on economic and political discourses. Yet, rather surprisingly, it has not had the same impact on philosophy or, more generally, theory, which has tended to address it only marginally, if at all. Only in recent years has analytic philosophy embraced Darwinian biological models as paradigms of mind,[2] and it is even more rare to find philosophers from the Continental tradition invested in exploring the philosophical implications of Darwin's work.

Instead of examining the scientific development and elaboration of Darwinism through the nineteenth and twentieth centuries—a fascinating project but well beyond the scope of this book[3]—I concentrate in this and in the next two chapters on Darwin's own writings on natural and sexual selection, and on elaborating his understanding of evolution as the emergence in time of biological innovation and surprise. It is my claim that although there are acknowledged and well-recognized gaps and points of unclarity in Darwin's understanding—most notably, his self-avowed ignorance of the mechanisms of inheritance, published in its earliest and most speculative form by Gregor Mendel in 1863, only a few years after the publication of the first edition of Darwin's *Origin of Species* (1859/1996)—his account of the development of species, including the descent of man, provides a powerful and fundamentally plausible and suitably complex understanding both of the genesis of (primitive) life from the complexities of matter and of the growing elaboration, adaptation, and specialization of organisms to their stable or changing life conditions. Darwin's work is of direct relevance to feminist concerns and indeed is a commonly elided assumption of much feminist work, even as it tends to be identified with patriarchal privilege. Darwin develops an account of a real that is an open and generative force of self-organization and growing complexity, a dynamic real that has features of its own which, rather than simply exhibit stasis, a fixed essence or unchanging characteristics, are more readily understood in terms of active vectors of change. Darwin managed to make this dynamism, this imperative to change, the center of his understanding of life itself and the very debt that life owes to the enabling obstacle that is organized matter. This dynamism of life is the condition of not only cultural existence but also cultural resistance. While presenting an ontology of life, Darwin also provokes a concern with the possibilities of becoming, and becoming-other, inherent in culture, which are also the basic concerns of feminist and other political and social activists.

What Is "Origin"?

The question of origins, and of originality, is paradoxically not only the buried center of Darwin's concept of the evolution of species, it is also one of the critical historical questions directed to Darwin's own discourse. It is today a truism that Darwin's *Origin of Species* (hereafter OS) precisely refuses to deal with the question of the origin of species![4] It is also well recognized that Charles Darwin is not really the originator of the theory of natural selection, of modification with descent, or the struggle for existence, though his name is now singularly associated with a new discipline, or a new approach to an old discipline, bringing a mass of scientific information, a vast repertoire of empirical observations, together to produce an ever more credible and carefully detailed account. One of Darwin's first critics, Samuel Butler, in *Luck or Cunning?* (1887), charged Darwin with refusing to acknowledge the origins of the theory of natural selection by refusing to admit his utilization of already existing sources, particularly the work of Georges Buffon, Jean-Baptiste Lamarck, Robert Chambers, and, above all, Erasmus Darwin, Charles's paternal grandfather. It is one of the quirks of history that many of these ideas were developed in the first instance by the elder Darwin (1731–1802), a prominent doctor, naturalist, inventor of quite sophisticated machines (including various machines of locomotion anticipating the steam engine), as well as a political activist and free thinker (who struggled, among other things, against the institution of slavery and for the humane treatment of the insane, abstaining from both alcohol and Christianity).[5] The origins of the theory of the origin of species is itself indeterminate and impossible to pin down precisely.[6] However, I am less interested in the genealogy of Charles Darwin's ideas than in their form and structure, in their philosophical and ontological investments, that is, in understanding how his account of evolution works, what limits and problems it might have, what this account adds to the ways we understand time, matter, and life, and ultimately, how these explain or limit cultural life and provide us with more complex ways of understanding politics and political change.

The question of how to represent or understand the origin of species is intimately bound up with the question of how to understand the identity, or unity, of the object of biological and historical investigation. This is among the most complex and underdiscussed elements of Darwinism, the point where Darwin's own account uncannily anticipates Derridean *différance.* What is the minimal unit, the scientific object, of investigation: the individual, the group, the species, or life in its generality? How species develop and undergo modification over the passage of time is closely linked with the criteria of differentiation between one group and another closely allied with it. What differentiates one species from another? How do we tell where one species ends and another begins? How small or large must the differences between them be for us to designate the emergence of new species from already existing ones? These are the questions any science, at its inception, must ask in order to attain scientific status: What, in the clearest terms, is the object of analysis, and how can this object be decomposed into its most elementary parts? In attempting to devise workable (and necessarily anti-essentialist) answers to these questions, Darwin inadvertently introduces a fundamental indeterminacy into the largely Newtonian framework he aspired to transpose into the field of natural history: the impossibility of either exact prediction or even precise calculation or designation, the seeking of tendencies rather than individual causes, of broad principles rather than universal laws. Darwin introduced a new understanding of what science must be to be adequate to the reality of life itself, which has no real units, no agreed upon boundaries or clear-cut objects, and to the reality of time and change that it entails. This differentiated his understanding of natural selection from that of his contemporaries and predecessors: such a science could not take the ready-made or pregiven unity of individuals or classes for granted but had to understand how any provisional unity and cohesion derives from the oscillations and vacillations of difference. The origin can be nothing but a difference!

Darwin provides little discussion and no real explanation of the *origin* of species. He analyzes only the *descent*, the genealogy, the historical movement (for we cannot even call it progress) of species, the movement from an earlier to a later form, a movement that presupposes an origin that it cannot explain, which perhaps is not an origin except in retrospect. The question of origins, the emergence of the first forms of life in simple strings of proteins from nonliving chemical mixtures, has recently become a focal point of evolutionary conjecture, even though Darwin did indulge in some brief and undeveloped speculation on this question toward the end of *Origin:*[7]

> I believe that animals are descended from at most only four or five progenitors, and plants from an equal or lesser number.

> Analogy would lead me one step farther, namely, to the belief that all animals and plants are descended from some one prototype. But analogy may be a deceitful guide. Nevertheless all living things have much in common . . . so that all organisms start from a common origin. (OS 642)

Though Darwin seems to be reluctant to address the highly speculative question of origins, and though he lacked any scientific evidence to aid in these speculations, he does hypothesize that it may be the case that "all the organic beings which have ever lived on this earth may be descended from some one primordial form" (643), but he never goes further than to suggest that this is highly conjectural and, in some sense, ultimately irrelevant. It does not matter for the evolutionary project *how* life arose, as long as when it did, it conformed to the principles of individual variation and natural selection. Moreover, though we may not understand how or even when this transformation from the inorganic to the organic takes place, we can be assured that it did take place, for our own existence is proof of it. Darwin is extremely careful in *Origin* to provide arguments only for those claims for which he has amassed scientific evidence, but he does speculate in private correspondence to Hooker about this "moment" of emergence of life from nonlife: "But if (and oh what a big if) we could conceive in some warm little pond with all sorts of ammonia and phosphoric salts, light, heat, electricity and etc., present, that a protein compound was chemically formed, ready to undergo still more complex changes" (quoted in Depew and Weber 1997, 399–400).

This remains a question of enormous interest to contemporary Darwinians, to molecular biology, to ethological simulations, and especially to those working in the arena of artificial life, a-life (the modeling and simulation of life, usually very simplified representations of life, in open-ended computer programs), as well as to philosophy: Out of what raw materials and using what processes did the simplest forms of life emerge? What were the non-organic ingredients of the prebiotic soup out of which elementary life appeared? What is the point of conversion from chemical to biological components? How closely tied are biological life forms to the particular chemistry of those forms of life with which we are familiar? We, and all creatures on earth, are carbon-based life forms. The question contemporary a-life scientists ask is: Is life essentially tied to those accidental, carbon-based life forms we know today? Can there be a silicon-based or, say, a nitrogen-based life form? What, for example, are those open-ended computer programs that exhibit reproductive, regulative, and emergent properties, similar, at least in some respects, to other forms of life? Was the emergence of life highly improbable, a freak occurrence, unlikely to be repeated?[8] Or was it, immanent in the inorganic chemistry of the universe, a likely consequence of chemical potential?[9] Just what kind of chemical complexity and types of transformations are necessary to precondition or refigure life?

This is in part to ask the philosophical question: At what point and in what form does matter convert itself, through whatever chemical/informational reactions, into life, however simple? At what point is there a transformation from quantitative to qualitative? At what point does material or informational complexity become organic, at what point does matter become complex and coordinated enough to be considered living, and what material constraints exist on the processes of information in "living" systems? Is life materially tied to the biological forms that appear in the past and today? Is there a pattern, a network of information, that is substrate-neutral? Is life a pattern rather than a mode of materiality?[10]

Instead of a theory of genetic origin, or a theory of descent from original primordial ancestors, paradoxically and without much analysis by other commentators Darwin seems to produce a quite peculiar, and thoroughly postmodern, account of origin. Origin is neither a divinely ordained beginning ex nihilo, a magical creation or gift, nor is it the result of an infinite, unbroken material and historical chain of organisms linked through descent, the two residual theological models, creationism and infinite or eternal existence. Origin is a consequence of human, or rather, scientific taxonomy, a function of language. Origin is a nominal question. What constitutes an origin depends on what we *call* a species, where we (arbitrarily or with particular purposes in mind) decide to draw the line between one group and another that resembles it, preexists it, or abides in close proximity with it. What we call a species depends on certain affinities and resemblances, as well as on differences and incompatibilities between different groups. A species is an arbitrarily chosen set of similarities that render other differences either marginal or insignificant. Species are a measure, an incalculable, non-numerical measure, of significant differences. The individuals constituting each species vary immensely from each other. When these variations exhibit some systematicity and resemblance, as well as significant or demarcatable differences between groups, we may be justified in describing the individuals thus associated as a variety; similarly, it is only if the variety has marked any significant differences from other varieties that it has the potential to develop into separate species, genera, or phyla. The origin of species can be understood as the discernable but noncalculable measure of degrees of difference between individuals and groups, a kind of biological *pure difference.*

The term "evolution" itself, derived from the Latin *e-volvere,* means "to roll out," "to unfold." It is a term Darwin himself avoids, preferring the phrase "natural selection," but it is perfect to capture this difference that is never based on a given unity but on a broad community-in-difference and common history, which could be understood as biological "memory," the present traces and supersessions of the past, the continuing inherence of a personal or prepersonal, individual or collective, past in the present.[11] We will return to this understanding of evolution as a durational unfolding, as the elaboration of difference, in future chapters. What evolves are not individuals or even species, which are forms of relative fixity or stability, but oscillations of difference (which underlie and make possible individuals and species) that can consolidate themselves, more or less temporarily, into cohesive groupings only to disperse and disappear or else reappear in other terms at different times.

There is no given mode of definition or genealogical method that could, without arbitrariness, provide clear-cut units to undertake the retrospective and reconstructive search for origins that any historical method implies. One could impose definitions a priori on species and varieties, as did the neoclassical Aristotelians such as Buffon and Cuvier, but then one would have to explain the generation of vast anomalies, and in doing so, one would lose sight of what justifiably enables individuals to be grouped into categories or types: "No criterion can possibly be given by which variable forms, local forms, subspecies, and representative species can be recognised" (OS 72). Darwin elaborates: "Certainly no clear line of demarcation has as yet been drawn between species and sub-species—that is, the forms which in the opinion of some naturalists come very near to, but do not quite arrive at, the rank of species; or, again, between sub-species and well-marked varieties, or between lesser varieties and individual differences. These differences blend into each other by an insensible series; and a series impresses the mind with the idea of an actual passage" (77).

The differences between individuals do not thereby generate differences in kind between species. Individual differences form continua, whose divisions remain relatively arbitrary, contingent on the pragmatic purposes of the division. Differences that "blend into each other by an insensible series" are differences of degree rather than of kind.[12] Yet, it is through the continual production of degrees of difference, over long enough periods of time, that kinds of difference will emerge. The differences between individuals, if they are significant, if they *make* a difference that is pronounced enough, produce varieties, which, if their differences are significant, generate the capacity to be categorized together as species and differentiated from other species. Darwin makes it clear that it is not simply differences, or even differences mingled with modes of repetition/reproduction, that constitute the basic categories of life and of species, but rather, modalities, types, or *degrees* of difference: it is "the passage from one stage of difference to another" (OS 78) that is central to the operations of natural selection, the movement of differentiation that, gradual and possibly imperceptible as it is, marks off varying degrees of differentiation of parental from offspring generations. Variations are crucial for the breadth, density, and longevity of species, but these variations are the very mechanism by which differences in kind are born and species disappear. However, what constitutes a variation and what remains simply an individual modification within a species is not just qualitative, for the quantitative plays a significant role in the movement from variety to species: a variety is considered a variety only insofar as its numbers have not yet surpassed those in the species of which it is a variation:

> If a variety were to flourish so as to exceed in numbers the parent species, it would then rank as the species, and the species as the variety; or it might come to supplant and exterminate the parent species; or both might co-exist, and both rank as independent species . . .
>
> From these remarks, it will be seen that I look at the term species as one arbitrarily given, for the sake of convenience, to a set of individuals closely resembling each other, and that it does not essentially differ from the term variety, which is given to less distinct and more fluctuating forms. The term variety, again, in comparison with mere individual differences, is also applied arbitrarily, for convenience' sake. (78–79)

In other words, Darwin's objects of analysis are distinctive or significant differences, differences whose divergence can be understood only differentially, that is, retrospectively and comparatively—historically—rather than in terms of any unchanging characteristics, content, essence, function, or morphological feature. There is no origin of species because there is no unity from which descent is derived, only types, variations of differences and types of reproduction and descent, both of which must be assumed from the start. This structure of variation or differentiation, as some contemporary biochemists hypothesize, may in fact characterize the status of those prebiotic elements whose status "on the edge of chaos" prepares for their possible transformation into primitive proteins (Prigogine and Stengers 1984; Kaufman 1993).

In the beginning, if it makes any sense to talk of a beginning, there were differences, in all likelihood chemical differences, variations in compounds, in different geographical and climatological contexts. These chemical differences, under some unknown and perhaps unknowable conditions, were transformed or transformed themselves into simple organic proteins, whose structure provided some means of reproduction. Life "began." This origin, as much fable as strategic assumption, is not only obscure, conceivable only through abstract reconstruction or speculative genealogy, but is in a certain sense impossible to understand as a locatable or knowable entity, a definite point in time, a single chemical reaction, for it is an origin "that is not one," that is always already implicated in multiplicity or difference, in a constellation of transformations, an event that imperceptibly affects everything.

The "Origin" of Language

Darwin returns again and again, without resolving it, to the question of origin and descent. The most significant and well-developed analogy he draws in *Origin* to explain biological descent or genealogy is with the history of languages, a sustained homology between natural history and philology. And, as in the case of species, the origin of languages remains obscure and possibly incalculable. It, too, involves some arbitrary dividing line separating the primitive, nonlinguistic utterance of sounds in protohumans from the vast and immensely variable systems, all fully formed and perfectly functional, as Darwin recognized, that we understand as language. This relation is a gradual one, a movement of continuous development without sharp boundaries or a clear point of origin, without unity or a single moment of emergence: "We know hardly anything about the origin or history of any of our domestic breeds. But in fact a breed, like a dialect of a language, can hardly be said to have a distinct origin" (OS 59).

Darwin figures the functioning of already existing and past species on the model of already existing and past languages, which develop only through the accumulation of minute variations that are, or are not, successfully propagated, and whose historical existence is figured on both their capacity to generate variations and their capacity to keep variations integrated and cohesive, in consonance with their preexisting forms even while developing out of them.[13] This inherent drift or movement of change within language, when it becomes further differentiated and specified, eventually results in the production of new dialects, which may eventually, over time, further differentiate themselves to become distinct languages.

The same problems regarding the origin of species face any account of the origin of languages, and the same inherent indeterminacy regarding the unit of analysis—the word, the sentence, the text, a dialect, a language—haunts linguistics as it does biology. If these units are not simply given to observation, if they are sub- or superobservational, that is, if they are methodological presuppositions that depend on how we define our terms, "species" or "dialect," then it is no longer clear how these units are to be agreed upon by researchers, what criteria one may use to secure agreement, and how scientific methods may develop to analyze these modes of continuous linguistic and biological variation.

What species and languages share is a reliance on the concept of genealogy. Both are fundamentally temporal processes, capable only of retrospective rather than prospective analysis, which involve the hypothetical reconstruction of a past that has left only fragmented and decaying traces or remnants, modes in which order, timing, and precedence are irreducible factors, in which historicity and the movement forward of time are necessary considerations. As we shall see, this will lead, in the late nineteenth and early twentieth centuries, to a reversal of the split between physics and biology, for it is only when physics moves away from the model of atomistic causation and toward a probabilistic model that the "arrow of time" becomes a relevant and irreducible principle of thermodynamics. Species and languages are defined not by their characteristics, their material, that is, morphological and phonological, features, but through their relations of descent, where they have come from, and on broad principles of change gleaned from the historical forms of change they have already undergone. Darwin describes them as "genealogical" systems, systems of pedigree, filiation, or inheritance. This signals their broad resemblance to other historically bound systems—to economies, to cultural systems that are equally embedded in the movement of history: "Thus, the natural system is genealogical in its arrangement, like a pedigree: but the amount of modification which the different groups have undergone has to be expressed by ranking them under different so-called genera, sub-families, families, sections, orders and classes" (OS 562).

Because both biology and language are systems of reproduction with minor variation, where the degree of difference from the "parental" forms, living or linguistic, marks any variation or innovation and where high degrees or intensities of difference are not tolerable but function as deformations, species and languages retain both a certain family resemblance to and a certain degree of difference from their parental forms. The resemblance between languages and the development of species is not, of course, random or contingent, for, as Darwin suggests, if we could somehow devise a history of all of human existence over time, we would have precisely the genealogy of languages so sought after in philology. The history of languages *is* the history of the movement of populations and the transformation effected on these populations by their changing environments and interactions with other language-speaking populations. It is not surprising, then, that the history of languages exhibits congruences with the development of species: its own history is intimately embedded in the relatively recent history of human, and thus cultural, development.[14] The development of language, as part of the processes of natural selection, is subject to the same evolutionary pressures, and its genealogy is contingent on that of its human speakers. The selective pressures on languages are the same as those on particular populations and cultures:

> It may be worth while to illustrate this view of classification, by taking the case of languages. If we possessed a perfect pedigree of mankind, a genealogical arrangement of the races of man would afford the best classification of the various languages now spoken throughout the world; and if all extinct languages, and all intermediate and slowly changing dialects, were to be included, such an arrangement would be the only possible one. Yet it might be that some ancient languages had altered very little and had given rise to few new languages, whilst others had altered much owing to the spreading, isolation and state of civilisation of the several co-descended races, and had thus given rise to many new dialects and languages. The various degrees of difference between the languages of the same stock would have to be expressed by groups subordinate to groups; but the proper or even the only possible arrangement would still be genealogical; and this would be strictly natural, as it would connect together all languages, extinct and recent, by the closest of affinities, and would give the filiation and origin of each tongue. (OS 562–563)

As with biological descent, Darwin posits a model of temporal fanning: beginning from a small, localized, unspecifiable time and place, languages, like living forms, develop over time (and geography) by differentiating and proliferating themselves, spreading themselves through wider territories and, in the process, encountering other languages, assimilating or being assimilated by them, transforming themselves, evolving. The growth of languages over time is the development of ever more elaborately bifurcating dialects, many becoming extinct in the process and only some leading to languages as we know them today. The analogy with which Darwin begins becomes a homology, an elaboration or ramification of biological evolution itself: the development of language is not just *like* evolution, it *is* evolution. Languages, like forms of organic existence, have their own ways of developing over time, their own broad principles, probabilities, and preferences, their own "logical" or internal force, their impulses to proliferation, which are confronted with the forces of natural selection, that is, with ongoing use in the context of unexpected encounters with hostile or beneficial (linguistic and social) forces.

Darwin's understanding of linguistics is remarkably astute, given the emphasis on comparative philology in his time. He is above all interested in language's capacity to sustain, even require, ongoing change while maintaining its systematicity and integration, its correlated growth, in the terms of Darwinian biology, at any given moment of time. In this, his understanding anticipates the integration of synchronic and diachronic linguistics, which had to wait until the early twentieth century and the writings of Swiss linguist Ferdinand de Saussure (1974). The link with Saussure's work is closer than it may seem; indeed, it would be surprising if Saussure wasn't in some way affected by the transformation in scientific models Darwin's work accomplished. In particular, at the origin or heart of both biological and linguistic models is pure difference, a difference without positive terms, an indeterminacy that replaces origin and that substitutes for the only identity of the units analyzed:

> The formation of different languages and of distinct species, and the proofs that both have developed through a gradual process, are curiously the same . . . We find in distinct languages striking homologies due to community of descent, and analogies due to a similar process of formation. The manner in which certain letters or sounds change when others change is very like correlated growth. We have in both cases the repudiation of parts, the effects of long-continued use, and so forth. The frequent presence of rudiments, both in languages and in species, is still more remarkable . . . Languages, like organic beings, can be classed in groups under groups; and they can be classed either naturally according to descent, or artificially by other characters. Dominant languages and dialects spread widely and lead to the gradual extinction of other tongues. A lan-

guage, like a species, when once extinct, never, as Sir C. Lyell remarks, reappears. The same language never has two birth-places. Distinct languages may be crossed or blended together. We see variability in every tongue, and new words are continually cropping up; but as there is a limit to the powers of the memory, single words, like whole languages, become extinct. As Max Müller has well remarked:—"A struggle for life is constantly going on amongst the words and grammatical forms in each language. The better, the shorter, the easier forms are constantly gaining the upper hand and they owe their success to their own inherent virtue." (Darwin, *Descent of Man*, 1:59–60)

Here, in *The Descent of Man* (hereafter DM), Darwin describes a series of striking congruences between species and languages:

1. Like species, languages submit themselves readily to different modes of classification, either artificial—in terms of resemblances, analogies, contiguities, and so on—or in terms of the more "natural" classification provided by the lineage of descent.

2. Changes at the level of individual words and phrases, like variations at the level of individual life forms, gradually and slowly perform the labor of converting variations and species or dialects and languages only if they succeed at all in affecting collective change. The agents of variation in both biological and linguistic systems are the relatively random changes that occur at the level of individual terms, which generate small but complex effects on entire systems, whether species or languages.

3. These individual changes are rarely isolated or unconnected with other indirect effects; it is unusual for a single change to occur without correlated effects and repercussions on the other terms (a correlated set of variations that direct change in multiple ways).

4. Both languages and species contain within their present forms rudiments or residues of their own previous historical forms, vestiges that inscribe their present form with the traces of their past, living forms of memory.

5. Languages are subjected, within certain limits, to exchange with other languages, to extend and magnify themselves, but beyond those limits, they become part of the struggle for survival, facing assimilation, extinction, or transformation.

6. Like species, languages face two kinds of pressure from the forces of natural selection: a pressure from the force of competition between terms within a given language, which leads to the abandonment of some terms and

the elevation of others, and thus to development and change within a single language, its adaptation to present usage; and a pressure from the force of other languages, which, through cultural, economic, and/or military privilege, or perhaps linguistic prevalence, leads to the extinction of some languages and the gradual development of new languages out of the remnants of surviving ones. Languages are subjected to precisely the same principles of individual variation and natural selection that are the motor for organic development.

7. Darwin suggests that, as is true for species, it is difficult to understand the concept of a "primitive" language or a language-in-the-making, a protolanguage, for every language is uniquely adapted, has survived for a length of time, and has thus had its use-value, its survival value, over the sign system it must have replaced. There may be "barbarous" cultures, but there is no such thing as a primitive language; just as with species, there may be simple or complex creatures, but the simpler ones are no less well adapted to their life situation than the most complex. Each articulates itself equally well: "A Crinoid sometimes consists of no less than 150,000 pieces of shell, all arranged with perfect symmetry in radiating lines; but a naturalist does not consider an animal of this kind as more perfect than a bilateral one with comparatively few parts, and with none of these alike, excepting on the opposite sides of the body. He justly considers the differentiation and specialisation of organs as the test of perfection. So with languages, the most symmetrical and complex ought not to be ranked above irregular, abbreviated, and bastardised languages, which have borrowed expressive words and useful forms of construction from various conquering or even conquered races" (*DM* 1: 61–62).

The study of languages in their diversity is best facilitated, not through evaluations of their internal complexity, their essential features, but in terms of both their history and their adaptedness, their functionality in the present. Ironically, philology is more "scientifically" developed, more historically elaborated, more linked in terms of genealogical affiliations than natural history had been. This fundamental undecidability, which can and has been resolved by fiat, about the notion of origin, the origin of species, the origin of languages, and about the units of investigation—whether higher-order systems (languages, families of languages/species, genera) or more localized and individualized groups (words, phrases, sentences/individuals, varieties, subspecies)—will pervade Darwinism and problematize its scientific value. There will be numerous problematic attempts to specify what counts as a decontextualized "fitness," or evolutionary success, for both languages and species, but it is precisely this indeterminacy, so central and acknowledged by Darwin, that will prove of major value in a Darwinian ontology of becoming.

The Political Economy of Nature

Darwin claims that three basic and closely linked principles, taken together, explain the contrary forces at play in the evolution of species: individual variation, the reproductive proliferation of individuals and species, and natural selection. The evolution of life and the evolution of language are possible only through the irreversible temporality of genealogy, which requires Darwin's three closely interrelated and codependent processes: first, an abundance of variation; second, mechanisms of indefinite, serial or recursive replication or reproduction, and thus the long-term magnification and elaboration of this variation; and third, criteria for the selection of differential fitness among competing individuals, varieties, and species. When put into dynamic interaction, these three processes provide an explanation of the dynamism, growth, and transformability of living systems, indeed the impossibility of stasis and mere reproduction, the impulse toward a future that is unknown in and uncontained by the present and its history.

It may be claimed that these three principles provide a more generalized explanation not only for the operation of biological and ecological, that is, natural systems, but also for the temporal operations of various social, economic, and cultural systems, systems that exhibit principles of self-organization that yield the sometimes gradual and sometimes surprising emergence of a nondirected or uncontrolled, bottom-up order. This suggestion has been the basis of a series of fascinating experiments in computer simulations, for example, of the stock exchange, population growth, war strategy, and other cultural scenarios.[15] It can even be argued, as many commentators have, that these evolutionary principles provide a more generalizable explanation of the logic or pragmatic structuring of cultural systems themselves, which, though clearly different in orientation and in some material components from biological systems, are nevertheless subjected to some of the same principles of organization and are clearly dependent on them.[16] Indeed, the principles of self-organization that seem to underlie and are presupposed by evolutionary theory may apply equally well to clearly inhuman systems, such as the weather, as they do to cultural, human systems, such as traffic. Weather and traffic are both systems in which there is tremendous financial investment in devising forms of control. Yet it remains unclear whether, as with biological systems, we must remain content with retrospective knowledge or can look forward to predictive knowledge. Weather and traffic are regulated by a few basic principles (in the case of the weather, for example, the intensity of air currents, the differential temperature between ascending and descending air movement, and the gradients of temperature between higher and lower regions of the atmosphere; in the case of traffic, the speed of drivers, the number of cars, the width and length of the road, the number of intersections) that attempt to structure the behavior of large numbers, sometimes even millions, of units or elements. Broad patterns of regulated behavior emerge, but under certain conditions, these systems tip into chaotic forms, creating upheaval and unpredictability.

Thus, for example, it has been argued with some plausibility that while Darwin derived a number of insights about the development of species through extrapolations from the political and economic writings of Adam Smith and Thomas Malthus, his account of natural selection may provide an explanation for economic history as readily as for natural history (Barrish 1991).[17] Darwin himself claimed that the theory of natural selection "is the doctrine of Malthus applied with manifold force to the whole animal and vegetable kingdom" (OS 63). The interlinkage of natural history, linguistics, and economics is not random, for each studies relatively self-contained but fundamentally open-ended systems that are temporally and geographically sensitive in their operations and are subjected to tendencies and probabilities rather than to laws. Under certain conditions, each can tumble into chaotic processes under which systems break down or new forms emerge. Each studies ever-changing systems, where history provides, not a definite goal or direction, but open-ended resources and constraints, and in each the systematicity of the system studied depends on an exchange relation with other systems and with forces of asystematicity or destabilization. Each is a temporal system, directed always forward, becoming always more complicated as time progresses, but without being able to predict where or how. As temporal systems, all that remains clear, if that, are elements, residues, of the past that remain in the present and set broad goals and constraints on the future.[18]

Malthus's (1798/1976) theory, in brief, is that as all living populations have a tendency to increase their numbers at a "geometrical," that is, exponential, rate, and as natural resources, territory, and so on remain limited and develop only at a linear rate, in the long run there is inevitably a situation of increased competition, and thus a struggle for life and death among competitors over increasingly scarce resources. It paints a pessimistic portrait of the inevitability of scarcity, struggle, and extinction, yet the struggle for existence also implies that under certain conditions, particular individuals and groups, through their own resourcefulness, through individual characteristics, quirks, or inventions, may find a specific niche to exploit, a way of providing for their own subsistence, until a more able competitor arises. Malthus was himself strongly influenced by Newton and Newtonian models of law and system. He, too, conceptualized a fundamentally closed system with two broad parameters or operative principles: growing populations and dwindling resources. The relation of these components is governed by discoverable laws: organisms tend to maximize their offspring up to the limit of available resources, for resources tend to grow at a slower rate than organisms reproduce. Equilibrium between demand and supply is continually restored through the force of scarcity, which, some have argued, functions in a law-like fashion for Malthus as an analogue of Newtonian gravity (Gould 1989; Depew and Weber 1997, 120). These systems, which in human populations we can understand as economies, closed systems, function like Newtonian steam engines, whose internal limits can be understood a priori according to the laws governing the conservation of energy: as the economy expands and populations grow, resources (energy) are used up, slowly, relentlessly directing the system toward a terminal state ahead, a state of the long-term depletion of attainable resources, a kind of economic heat death equivalent to and understood on the model of the entropic unwinding of the universe itself.

Malthus's proposal provided Darwin with the idea of the effects of pressure or impetus, the conditions under which natural selection would be able to select the better adapted from the less well adapted, to reproduce more numerously than their competitors and to enter into a process of refinement of their adaptedness to their niches through the exertion of other selective pressures. The impetus for such change is precisely the increasing pressure of competition between individuals and species for ever more rare resources. Increasing differences between these groups is the result of these selective, that is, competitive, pressures, the force that selection places on variable species to explore and exploit new resources and territories, new niches, and to become differentiated relative to each other and adapted relative to those environmental niches in the process. The pressures of population growth not only weed out the less adapted, those less able to subsist; they also impose an impetus on those who survive to produce increasing resourcefulness, to stimulate colonization, economic initiative, and the exploitation of new markets and new resources. Darwin set Malthus to work, not in the realm of economic production and social existence, but in the teeming world of species as a whole.[19]

Malthus was perhaps the most direct force of inspiration for Darwin's concept of species as competing agents, but he was also strongly influenced by Malthus's precursor, Adam Smith, and Smith's friend, David Ricardo (1817). This cluster of economic texts approaches the social and the economic as systems of self-regulating forces, where the human is not a sovereign agent, but one site of selective pressure, one point in a broader, nonhuman or megahuman system. In *An Inquiry into the Nature and Causes of the Wealth of Nations*, Smith (1776) argued for the "invisible hand" of the market as the rationalizing motor regulating individual self-interest and stabilizing economic relations, generating autonomous rules and principles for the agents in the market and for the resources with which they deal.[20] Smith was concerned to provide something like a Newtonian understanding of the laws and forces producing and regulating wealth, the laws of supply and demand, which regulate and act as a force of selection and constraint on self-interest.[21] The pressure for invention, for innovative production on the part of individual economic agents, Smith argued, is the engine of economic progress insofar as it provides the impetus for mass production and for diversified life in the market. In accordance with the dictates of self-interest, each economic agent in a market tries to produce commodities as cheaply as possible and to sell them for as much profit as possible. The market is driven by a formula, a maxim: buy (commodities) low and sell high. The principles by which the market is driven—what commodities are made, how many are produced, what price they sell for, what profit they accumulate—are calculable from these market-produced and -driven, supervening principles. The market is an equilibrating system, which adjusts itself according to the inputs, the sources for the production of wealth, and regulates the outputs, commodities, according to its own emerging and elaborating logic.

For Smith, value is not found in the natural wealth of the earth—its accidental and uncontrollable fertility, which is liable to become more and more depleted over time—but in labor and only labor, the energetic input of producers, which is, as it were, conserved, like Newtonian energy, because it is stored, traded, consumed, and replenished in the form of commodities. Value is produced through an ever more efficient utilization of labor in relation to one's competitors, who in turn must produce more value through their own growing efficiencies or else face economic extinction. The market spontaneously creates its own balance between supply and demand, generating "natural" pricing for commodities—what the market will accept—and establishing an arena for competition among its various agents, where it proliferates ever growing efficiency in its own terms. This efficiency is most clearly manifest insofar as the market provides greater and greater incentives to divide and diversify labor and production, to search for, find, and exploit increasingly smaller and more specialized niches of consumers. What Malthus understood in relation to populations, Smith saw in the notion of an autonomous economy, a machine at work that forms its own natural systematicity if it remains untrammeled by external factors, especially external political or military forces. Like the current self-representation of the stock market (which nonetheless does have external constraints in the form of federal agencies that regulate it legally; the Federal Reserve, which periodically regulates monetary policy and interest rates; foreign exchange to indirectly intervene into its operations; and federal and state governments, which periodically adjust their tax laws to stimulate or dampen an economy), Smith's free market is a self-generating, self-regulating, dynamic machine, a mega-organism that, like the astronomical forces that concerned Galileo, exhibits a purity of operations, though unlike Galileo's system, it never repeats itself but always becomes more complex, more elaborate, more detailed and specialized.

Like Newton before him (in the field of physics) and Darwin after him (in the realm of biology), Smith believed that he had discovered the inherent laws regulating the closed system, in his case, the system that is wealth. Through the division of labor, a minimum number of resources will be utilized through a maximization of diversified functions. It is hardly surprising, given this idea of the dynamic, autonomously adjusting, self-regulating nature of the economic order represented in Smith's picture, that Darwin believed he could discover at least some of the laws, or, more accurately, principles by which the self-regulating systematicity of the natural order, of life and its cultural outcrops—language and economic production and consumption—function.

This is what the theory of natural selection seeks to provide, though in the process, Darwin's complex view brings out latent complexities and problems for free market economic theorists who followed Smith. Ricardo (1817), for example, in *Principles of Political Economy and Taxation*, points to the long-term falling rate of profit as the economy moves closer to a steady state. There are inherent limits to newly exploitable resources, and hence there is a necessary tendency, in the long run, to falling rates of profit. The capitalist's sole source of profit is the gap between the value that the worker creates and the financial compensation the worker receives. It is only through the creation and exploitation of new markets, which can remain new only for a relatively short period of time, that this fall can be temporarily halted. Karl Marx (1974) in *Capital* made it clear that as the economy expands, it also necessarily moves closer to a terminal state as the rift between the minimization of the laborer's wages and the maximization of the capitalist's profit widens and class antagonism and eventually warfare result.

Many have accused Darwin of blindly succumbing to the incipient ideology of bourgeois individualism, and thus of providing unconscious support for the growing influence of free market thinking. This may be too simplistic an understanding to be adequate to the subtlety and complexity of his work. Just as there are affinities and points of congruence between Darwin's understanding of natural selection and liberal political economy, so too there are close affinities with Marxism and other radical economic models, indeed, with virtually *all* economic models. This may have to do with the fact that, once the economy is considered a system, it comes to resemble in certain respects the very notion of system provided by life itself. The very model for systematicity may be provided by the organism. There are, for example, close resemblances between Darwin's understanding of individual variation and Marx's understanding of labor. Labor is that which is immensely individually variable, being closely linked to the innate particularities and learned skills of the laboring body and to the commodities that come to embody that labor. Individual labor becomes embedded in the use-value of the commodity, and the use-value is the means by which individual acts of labor become functional in systems of surplus-value and in the accumulation of capital. The specificity of labor becomes homogenized into a system in which the commodity produced by that labor gains its value outside of itself, by being made equal to something else (its exchange-value). Through the mirroring of labor in the form or body of the commodity, through its embodiment as use-value, which is then homogenized through exchange-value, differences in labor are ordered into systems of hierarchical structure that Marx explicitly models on biological categories: "Like the biological classification he explicitly models it on ('forms of useful labor . . . differ in order, genus, species and variety'), the division of labor as Marx conceives it does not simply refer to labor being partitioned into definable tasks, but to an order characterized by hierarchical and uneven levels of difference among its elements" (Barrish 1991, 437).

For Darwin, too, those random variations at the level of the individual are transformed into the hierarchized differences that constitute a natural, that is, genealogical or historical, taxonomy, and it is only their place within this taxonomy that provides the frame in which these variations can be understood as variations. What Marx made clear was that the economy cannot be regarded as a closed system, self-contained and autonomous, even if it contained within itself its own historical inevitabilities: it is also fundamentally linked to and central in the political and social arrangements of a culture. It is not a Newtonian closed system of self-contained forces but an open system in which the economy "leaks" into political and cultural, as well as biological, life.

The interlinkage between biology and economics—the study of living systems and the study of the production, circulation, and exchange of objects or commodities—is not fortuitous insofar as the two disciplines have long borrowed models, images, metaphors, and techniques from each other. Nor, as I argued in the previous section, is the link with linguistics and the systematic structure of languages accidental or contingent. In the aspiration to a scientific study of living phenomena, phenomena that actively transform themselves over time, whether these are biological, economic, or linguistic, each of these branches of knowledge has utilized and developed itself in relation to conceptual and methodological frameworks derived from the other two. In the early nineteenth century, each was inspired to provide a Newtonian reading of its objects of investigation, constraining each to understanding its object as a closed, given system, based on an atomism of its given units, the law-like nature of its operations, and to a calculable ratio of forces, providing criteria for success or failure. Each discipline underwent similar transformations provoked by perceived limitations in Newton's static ontology, the transformation in the physical sciences over a century ago from atomistic causation to population thermodynamics and probability theory, a shift from the vantage point of individuals to that of overarching, large-scale patterns. Even greater transformations of these three disciplines may occur as the result of the more recent scientific orientation to far-from-equilibrium systems and the logic of self-organization or complexity they entail.[22]

Darwin seems to intuit a close affinity between the open systematicity of linguistic and economic systems on the one hand, and biological systems on the other. This affinity is not simply borrowed; it is not merely the result of the fact that linguistic and economic systems emerge as the result of the elaboration of biological systems. Rather, unlike many asystematic elements of cultural and biological life, as far as language and economic production are concerned, a pattern of value emerges that comes to privilege, to select, some individual living forms, individual words and linguistic rules, or particular commodities rather than others. These systems are not human products but are *inhuman*: systems functioning beyond or above the control of their participants, systems that, as much as biological processes, form and produce their subjects.

We must now turn in some detail to Darwin's understanding of natural selection and its reliance on open unpredictability in both animal and human evolution. Then, in chapter 3, we examine how Darwin uses these principles to explain the evolution of the sexual and racial differences in the human.

2. BIOLOGICAL DIFFERENCE

Darwin claims that the dynamics of natural selection require individual differences, differences that can be "observed in the individuals of the same species inhabiting the same confined locality" (OS 67). The ongoing production of individual differences is the internal motor, the "vitalist impetus" of all of life. Life is a tendency to self-organization that ensures that living individuals, in ways unrecognized by Darwin and still relatively obscure today, vary from their parents and each other, not through the blending of parental characteristics, as Darwin presupposed, but through the complex processes of genetic selection, meiotic division, and recombination that characterize embryological development. These perceptual or motor differences are the raw materials of natural selection. These slight and usually minor individual variations have unknown causes. However, Darwin's ignorance of the principles of heritability did not compromise his understanding of evolutionary processes, in part because he did not seek out the causes of individual variation but only their effects. It is also significant that although some contemporary gene-centered research is attempting causal explanations of individual variation, the astronomical number of relevant genetic terms, the complexity of their rules of combination, and the impossibility of any one-to-one mapping of genes to phenotypical characteristics make this an unrealizable project, even given an accurate map of the genome. It is for this reason, among others, that genetic research is statistical or probabilistic rather than deterministic and is likely to remain this way. It analyzes tendencies and orientations rather than causal linkages, which means that it generalizes about individuals rather than analyzes them in their detail and specificity. It is this that enabled evolutionary theory to remain agnostic relative to, and independent of, the particular theories of genetics to be later developed. And this agnosticism is what facilitates the project of "grand synthesis" of evolutionary theory with the projects of genetics, as well as enabling it to be problematized, as it has been:[1] "We are profoundly ignorant of the cause of each variation. We are far too ignorant to speculate on the relative importance of the several known factors; and I have made these remarks only to show that . . . we ought not to lay too much stress on our ignorance of the precise causes of the slight analogous differences between species . . . The laws governing inheritance are for the most part unknown. No one can say why the same peculiarity in different individuals of the same species, or in different species, is sometimes inherited and sometimes not so" (31–32).

The Heritability of Variation

It remains true today that, although we have a clearer understanding of genetic structure and its coded nature, we are still unable to predict which phenotypical effects will result from which genetic causes. We still have no genetic explanation of living particularity. In a later work, Darwin (1885, 1:236) elaborates: "I have spoken of selection as the paramount power, yet its action absolutely depends on what we in our ignorance call spontaneous or accidental variability."

What genetics is still unable to explain is a direct causal connection between a genetic element—a single gene, a single chromosome, or even a string of them—and a singular phenotypic characteristic, "the cause of each variation." Darwin's acceptance of both the impossibility and the dispensability of any causal explanation was central to his reformulation of biology: from now on, what causes a change is of less interest than the antecedents, the ancestors, of that change. Randomness is introduced into the dimension of heritability. If it is clear that all inherited material comes from one's parents, it remains unpredictable which genetic characters will be selected and combined, which will function as dominant or recessive, in short, what particular kind of individual will be formed, until the individual is formed. Then retroactive analysis—the pointing out of resemblances, the recognition of continuities, the activity of genetic screening, the scanning of the genome—can begin. This remains as true today as in Darwin's time.[2] There is a fundamental randomness about the particular, detailed chromosomal structure of the individual, but even more significant, there is a fundamental randomness of individual variation *relative* to natural selection. They function independently of each other, variations being generated independent of adaptedness, generated, so it seems, for their own sake. When Darwin stresses that although many of these individual variations refer to minor or peripheral characteristics, features that may be neutral with respect to evolutionary success or fitness, there is evidence to suggest that the most significant, even defining characteristics of species also emerge through random variation. Individual differences, manifested in some way or other, either visibly or in terms of the acquisition and use of slightly different functions or modes of activity, are the object of natural selection, what the selective pressures of the environment single out and work on: "No one supposes that all the individuals of the same species are cast in the same actual mould. These individual differences are of the highest importance for us, for they are often inherited, as must be familiar to every one; and they thus afford materials for natural selection to act on and accumulate in the same manner as man accumulates in any given direction individual differences in his domesticated productions" (OS 67).

Darwin wants to link the most significant differences that constitute species, subspecies, and varieties, and the associated problem of the origin of species, to the differences between individuals. Indeed, his individualism is fundamentally linked to his anti-essentialism.[3] Species cannot be readily defined, for they are not constituted by essential features, abilities, or forms: species are nothing but the aggregation of interbreeding individuals who share a common descent. Many biologists before Darwin recognized that species varied, but such variation was regarded only in limited terms: species, like celestial bodies, underwent cycles of growth and reproduction, and this type of systematic and predictable variation, as well as those systematic differences that constitute the sexual bifurcation of many species, could well be accommodated in a typology rather than a genealogy of species. When Darwin construed species as a post hoc aggregation of individuals, what required explanation was no longer the possibility of individual variation but the converse, the long-term, relative stability of the characteristics attributable to species which nonetheless included widely varying individuals. The question is converted from How can individuals vary so widely? to How can species maintain their identity and cohesion over time? In effect, Darwin dynamized and historicized species through linking them to the continuous variation produced through individuals, developing an anti-essentialism of species only by propounding an essentialism of individuals (an essentialism that itself is being displaced, perhaps even deconstructed, through the understanding of individuals as the products of the differentiations of genetic and environmental factors that are each themselves differential rather than essential in their functioning).

It is the *degree of difference* between individuals that constitutes the possibility of varieties, and a greater degree of difference that constitutes the possibility of subspecies, species, and genera. It is this gradation of degrees of difference, in fact, rather than the individuals differentiated, that is the object of natural selection: "The passage from one stage of difference to another may, in many cases, be the simple result of the nature of the organism and of the different physical conditions to which it has been exposed; but with respect to the more important adaptive characters, the passage from one stage of difference to another, may be safely attributed to the cumulative action of natural selection . . . and to the effects of the increased use or disuse of parts. A well-marked variety may therefore be called an incipient species; but whether this belief is justifiable must be judged by the weight of the various facts" (OS 78).

Most individual differences under most sets of circumstances remain largely irrelevant to natural selection (though, as we will see, they may be more central to the operations of sexual selection); it is only under crisis conditions, when selective pressures are at their strongest, that they acquire the status of instruments of privilege and survival. Natural selection functions to select negatively for those characteristics that may prove harmful to the individual and its progeny and to provide a context in which individuals carrying those benign or benevolent characteristics are positively preferred over those individuals in which they are absent or less developed. These individual differences become relevant to the categories and forms of selection only when species are approaching a Malthusian state of maximum growth, where even slight or marginal advantages offer the individual some hope of survival over others:

> [Man] can neither originate varieties, nor prevent their occurrence; he can preserve and accumulate such as do occur. Unintentionally he exposes organic beings to new and changing conditions of life, and variability ensues; but similar changes of conditions might and do occur under nature. Let it be also borne in mind how infinitely complex and close-fitting are the mutual relations of all organic beings to each other and to their physical conditions of life, and consequently what infinitely varied diversities of structure might be of use to each other under changing conditions of life. Can it then be thought improbable, seeing that variations useful to man have undoubtedly occurred, that other variations useful in some way

> to each being in the great and complex battle of life, should occur in the course of many successive generations. If such do occur, can we doubt . . . that individuals having any advantage, however slight, over others, would have the best chance of surviving and of procreating their kind? On the other hand, we may feel sure that any variation in the least degree injurious would be rigidly destroyed. This preservation of favourable individual differences and variations, and the destruction of those which are injurious, I have called Natural Selection, or the Survival of the Fittest. Variations neither useful nor injurious would not be affected by natural selection, and would be left either a fluctuating element, as perhaps we see in certain polymorphic species, or would ultimately become fixed, owing to the nature of the organism and the nature of the conditions. (*OS* 108)

Following Malthus's formulation of the naturally exponential or geometrical rate of increase of reproducing individuals, it became clear to Darwin that, however slow the reproductive rates of various species, given enough time, any species will necessarily encounter scarcity and thus a hostile environment.[4] There is an invariable tendency to superabundance or excessiveness in the rates of reproduction and proliferation of species. Even if they merely reproduce their own numbers, species will eventually reach a limit of sustainable reproduction. This superabundance can be understood in more negative terms as the struggle for existence, which drives species and individuals to compete with each other for increasingly limited resources: "As more individuals are produced than can possibly survive, there must in every case be a struggle for existence, either one individual with another of the same species, or with the individuals of distinct species, or with the physical conditions of life . . . There is no exception to the rule that every organic being naturally increases at so high a rate that if not destroyed, the earth would soon be covered by the progeny of a single pair" (*OS* 54).

This teeming proliferation of individuals and species suggests that the greater the diversity, the more natural selection is able to take effect. If species reproduced only themselves or in ever-diminishing numbers, natural selection would be unable to weed out the less fit and provide space for the selection and profusion of the more fit, effectively preventing selection (*OS* 35). The production of greater and greater variability remains restrained by a number of factors. Although variation and proliferation are the very motors of the production of evolutionary change, there are nevertheless a series of limits on the degree of variability. For example, changes in life conditions may tend to increase variability (110). There is much dispute in current paleoanthropology regarding the fossil records and whether there was a vast proliferation of variation after the extinction of the dinosaurs ("punctuated equilibrium," in Gould's 1989 formulation), or whether there is a steady rate of variation that cataclysmic events highlight through the preservation of greater fossil numbers. The main issue of relevance here, in this philosophical analysis, is that the greater the rate of variability, the more material natural selection has to work on.

The range and scope of diversity or variability cannot be determined in advance, but it is significant that there are inherent, if unknown, limits to tolerable, that is, sustainable variation; "monstrosities," teratological variations, may be regularly produced, but only those that remain both viable and reproductively successful, and only those that attain some evolutionary advantage, either directly or indirectly, help sustain this proliferation. Darwin went to some lengths to make it clear that terata are only extremely rarely significant agents in evolutionary development, largely because natural selection does not function usually through cataclysmic changes or major upheavals, through saltation, but through the patient, slow, and meticulous acquisition of usually minor variations. It is common that minor changes in one characteristic may lead to major, potentially lethal changes in another. Individual variation is usually sustainable only when the mutations or changes are relatively minor and localized, though there are conditions under which localized or single variations can produce slow and cumulative changes in a population:

> It may be doubted whether sudden and considerable deviations of structure such as we occasionally see on our domestic productions, more especially with plants, are ever permanently propagated in a state of nature. Almost every part of every organic being is so beautifully related to its complex conditions of life that it seems as improbable that any part should have been suddenly produced perfect . . . If monstrous forms . . . ever do appear in a state of nature and are capable of reproduction (which is not always the case) as they occur rarely and singularly, their preservation would depend on unusually favourable circumstances. They would also, during the first and successive generations, cross with the ordinary form, and thus their abnormal character would almost inevitably be lost. (OS 66)

In themselves, individual variations have no value, though they are the positive motor and impetus of evolutionary change; it is only against the background of natural selection that these variations come to have value. Their value comes from their capacity to survive, to sustain themselves, and to reproduce. Like labor, or like phonological differences, individual variation must be placed into a system of selection to have value. Like origin, value itself is differential. Darwin places pure difference, pure biological difference, as the very matter of life itself: it is only differentiating, distinguishing, rendering more and more distinct, specializing and adapting that characterize life in its essence. Its essence is in differentiation, in making a difference.

Natural Selection

Taken together, the two principles of individual variation and the heritability of variation imply that, if there is a struggle for existence in circumstances where resources may be harsh or scarce, then any variation, however small and apparently insignificant, may provide an individual with advantages that may differentiate and privilege it relative to other individuals. Even minute variations may provide major advantages, especially in unexpected circumstances. Moreover, if individual variations are inherited, whatever small advantages were once bestowed on an individual may be amplified over time, creating exponentially increasing trends and directions of change. It is in this capacity for individual variation that Darwin locates the origin of species and genera. If individual variations are favored by natural selection and become a force in producing heritable characteristics, and if there is some separation, geographical or ecological, between such differentiated groups, conditions arise under which a new species, or several, can be considered to emerge from common ancestors.

If individual variation is the differentiating endogenous or internal factor in evolutionary emergence, then natural selection forms the exogenous, or external, differentiating element. Natural selection, or, in Alfred Russel Wallace's (1870) phrase, the "survival of the fittest," is the pressure exerted on individuals by both the environment, including other species, and members of the same species when they face conditions of scarcity and competition. As its name suggests, natural selection is the process, or rather processes (for it includes both artificial and sexual selection, as we will see below) that provide selective criteria that give significance and value to individual variations: "If . . . variations useful to any organic being do occur, assuredly individuals thus characterised will have the best chance of being preserved in the struggle for life; and from the strong principle of inheritance they will tend to produce offspring similarly characterised. This principle of preservation, I have called, for the sake of brevity, Natural Selection; and it leads to the improvement of each creature in relation to its organic and inorganic conditions" (OS 104–105).

Darwin describes natural selection as a "principle of preservation," but this preservation is quite ambiguous and multilayered. It preserves only those variations that can viably or even maximally function within the parameters or conditions in which the living being finds itself and that show some marked or significant advantage over competitors. The principle of preservation is the preservation of the fittest, of the most appropriate existences in given *and* changing circumstances. Fitness carries with it the notion of an openness to changing environments; it is not necessarily the best adapted to a fixed and unchanging context. (It is this understanding of an individual's or species' openness to changing circumstances that problematizes a common politically conservative reading of Darwin as apologist for the status quo: those with the most powerful positions of course want to justify their power through the justice of fitness. However, Darwin himself refuses to privilege any particular milieu, past, present, or future.) Fitness must be understood as an openness to the unknown, the capacity to withstand the unexpected as well as the predictable.[5] Through its sifting capacities, natural selection provides a negative mechanism, which functions to eliminate much of the proliferation generated by the hyperabundance of individual variation, indirectly sorting through the variations between individuals and species. It also provides a more positive productivity when it functions as the source of pressure on, or as an incitement to, those individuals and species that survive to even greater proliferation and divergence:[6] "[Natural selection] entails extinction; and how largely extinction has acted in the world's history, geology plainly declares. Natural selection, also, leads to divergence of character: for more living beings can be supported on the same area the more they diverge in structure, habits, and constitution, of which we see proof by looking to the inhabitants of any small spot or to naturalised production" (OS 105).

Mayr (1997) elaborates a point that Darwin merely hints at: although natural selection generally functions to weed out the less fit, it also proliferates, tests, and attunes the more fit; it induces genetic recombination, thus greater variation, thus more material on which it can undertake its selective work.[7] "When natural selection acts, step by step, to improve such a complex system as the genotype, it does not operate as a purely negative force, as many opponents of Darwinism maintained. It does not confine itself to the elimination of inferior gene combinations; rather, its most important contribution is to bring superior gene combinations together. It acts as a positive force that pays a premium for any contribution toward an improvement, however small" (45).

Natural selection functions only at the level of phenotype. Mayr argues that natural selection never functions on the level of genes, particularly never on the level of individual genes; rather, it operates only on the phenotype, whose characteristics are always structured by clusters or combinations of genes that function only in relation to environmental factors, as Oyama (2000b) affirms. It is not genetic fitness that is manifest in the individual's struggle for life, but phenotypic fitness: "Selection does not deal with single genes because its target is the phenotype of the entire individual. To assume that a given gene has a fixed selective value is an error because the contribution a gene makes to the fitness of an individual depends to a considerable extent on the composition of the genotype, that is, of the interaction of this gene with other genes" (Mayr 1997, 13).

If artificial selection—selection according to the criteria provided by human breeders of flora and fauna—functions to select characteristics that are visible and manifest, natural selection functions according to the usefulness, that is, the functional benefit, of phenotypic variations in all organs and functions: "Man can act only on external and visible characters: Nature, if I may be allowed to personify the natural preservation or survival of the fittest, cares nothing for appearances, except in so far as they are useful to any being. She can act on every internal organ, on every shade of constitutional difference, on the whole machinery of life. Man selects only for his own good: Nature only for that of the being which she tends. Every selected character is fully exercised by her, as is implied by the fact of their selection" (OS 111–112).

When selective pressures are at their greatest—and it is inevitable, given the exponential increase in populations, that this occurs sooner or later—the motor of natural selection functions in the most relentless fashion to sort out the more productive or beneficial differences from those less so. If individual variation may be considered the result of chance, the random combination of characteristics, then natural selection must be seen instead as the operation of a relentless, systematic orderliness, an automatic or, in Daniel Dennett's (1996) terms, algorithmic set of processes: "It may metaphorically be said that natural selection is daily and hourly scrutinising, throughout the world, the slightest variations; rejecting those that are bad, preserving and adding up all that are good; silently and insensibly working, *whenever and wherever opportunity offers*, at the improvement of each organic being in relation to its organic and inorganic conditions of life" (OS 112; emphasis in original).

Natural selection works most efficiently and relentlessly when the conditions of any species or group are approaching population saturation. However, natural selection cannot guarantee, first, that recessive or nonmanifest characteristics, even characteristics that may prove fatal in combination with similar recessive characteristics, are eliminated through its selective operations, for they play no manifest role in phenotype; and second, that the surviving individuals, those that survive long enough to reproduce themselves, are the most well-adapted of those born. In a passage that primarily notes the extravagance of natural selection and the destruction of vast potential forms of variation, Darwin also makes it clear that natural selection cannot guarantee the survival of only the fittest, but in fact settles for what is more pragmatic: the survival, through natural selection, of those who have survived random or accidental death, who not only are "fit" in terms of any skill or ability, but who simply survive through dumb luck. Natural selection, while it operates as an ordered and ordering network of processes, is in fact made up of nothing but thousands, millions of accidents, momentary events, that lead to the death of some, not because they were less well adapted but because they were, say, in the wrong place at the wrong time. These millions of small events, when charted statistically, produce a pattern of historically and geographically particular regularities that overall, statistically, but not in each individual case, select the more fit from the less fit:

> It may be well here to remark that with all beings there must be much fortuitous destruction, which can have little or no influence on the course of natural selection. For instance, a vast number of eggs or seeds are annually devoured, and these could be modified through natural selection only if they varied in some manner which protected them from their enemies. Yet many of these eggs or seeds would perhaps, if not destroyed, have yielded individuals better adapted to their conditions of life than any of those which happened to survive. So again a vast number of mature animals and plants, *whether or not they be the best adapted to their conditions*, must be annually destroyed by accidental causes which would not be in the least degree mitigated by certain changes of structure or constitution which would in other ways be beneficial to the species. (OS 116; emphasis added)

Darwin distinguishes here between two kinds of processes: random events, which destroy individuals independent of their relative fitness and which certainly have effects on the subsequent history of species that are indeterminate rather than related to adaptedness; and natural selection, which, though composed of random events, nevertheless functions as an overarching pattern that has the specific effect of preserving the stronger, the more fit, and eliminating the weaker, the less fit. This is a directed randomness, a randomness directed toward the selection of particular qualities, whatever they might be, which in some way positively advantages an individual. Is the first in fact an indirect or mediated form of the second? Is randomness, undirected randomness, not part of the very "testing" function of natural selection? The annual devouring of vast numbers of eggs and seeds is indeed part of the very operations of natural selection, and these seeds *may* have produced individuals better adapted than those that actually survived. The individuals who never developed into maturing adults—the evolutionary residue, those that leave no trace, no progeny—cannot simply be regarded as the losers, the inferior, in the evolutionary battle for adaptation. They remain the undeveloped, the latent, the recessive, a virtual forever unactualized, or perhaps actualized outside the traditional lines of genealogy. Two sorts of latent potentiality, virtuality, remain untouched by natural selection: that of the recessive characteristic, which can be selected, positively or negatively, only if it is combined with another recessive characteristic to produce an appropriate recessive-traited phenotype; and that which never had a chance to develop its capacity for variation, which was negatively selected before its fitness could be evaluated in more positive terms. These residues of selection cannot be simply conceived, in Hegelian terms, as that which is overcome or sublated, the negative that is transcended in a movement of the positive; they are not dialectical remnants but virtualities that remain unactualized, potentials unexpressed, forces redirected in their trajectory.

Natural selection works, not on the causes of individual variation, as Lamarck suggested, but on their effects. It cannot begin its relentless weeding out processes until individual variations manifest themselves phenotypically. There is a time lag, a delay, nick, or dislocation between variation and selection. Variation is the consequence of the still unknown relations of inheritance of genetic material from the parental generation; selection occurs with the descendants' struggle for existence. Lestienne (1998, 49) describes this as a causal rift, a separation between the causes of variation and the effects of variation, a separation produced not simply through the temporal delay between genotype and phenotype, but through their essential dissociation: "The rift's function is to create the necessary independence between the variation's causes and the selection's criteria." There may be unknown causes, too complicated for us to comprehend, but perhaps in principle understandable in the "laws" of inheritance, though this seems less and less likely; there are clearly highly particular causes for all the particular struggles undergone by members of each species. But these two causal series, if they can properly be called this, are discontinuous. Chance intervenes at their particular points of contact, where the forces of natural selection become intensified, where particular individuals are negatively selected for whatever momentary or long-term reason.[8]

Natural selection is the active, selective, and ever-transforming milieu of evolutionary change. It consists in the biological context of any existent, which is constituted largely, but not entirely, of the other living beings in their various interactions with each other. It also consists in the geographical, climatological, and highly specific material context for each existent, a nonliving context. These living and nonliving conditions enable natural selection to provide ever-changing criteria by which fitness can be measured and new variations produced. Natural selection is not simply a passive background or context in which individual variation unfolds, a landscape that merely positions and highlights the living being; rather, it is a dynamic force that sets goals, provides resources and incentives for the ever-inventive functioning of species in their self-proliferation and in their active transformation of this landscape of events.

The Descent of Mind

Many contemporary Darwinists want to find a scientific or rational mode of calculation, a *ratio*, for the multiplicity of small processes that make up natural selection. There is the desire, not just for the broad patterns that Darwin outlined, but for a detailed elaboration of the various processes that are at work in selection. Daniel Dennett (1996) in *Darwin's Dangerous Idea* may serve as a representative philosophical figure of this tendency. Dennett argues that, to be consistent with Darwin's dangerous idea, natural selection must assume a fundamental mindlessness to the operations of evolution. That is, one must fully accept that the "excellence of design," the apparently perfect adaptation of species to the specificities of their environment and for long-term survival, is the result of both serendipity or chance and the essentially blind and mindless system of selection that relentlessly weeds out and diminishes the effects and operations of the less adapted, and thus provides evolutionary advantages to the more and better adapted. As long as the time scale of evolutionary emergence is big enough, the mindless automatism of natural selection and the spontaneous production and inheritance of variation have time to ensure that experiments in living, as they might be called, living in a variety of environments under various conditions, produce maximal results from given resources, and those results in turn feed into those resources to actively transform them, which in turn transforms the stakes involved in selection in a productive, generative spiral.

Dennett argues that natural selection is an algorithmic process, a mindless, automatic, step-by-step set of procedures, like a recipe, which could, in principle, be carried out, or simulated, by a computer or machine, even though it might be composed of thousands, or millions, of small steps. Basing his speculations in part on Alan Turing's (1950, 1952) work on computation and biology,[9] Dennett (1996) claims that the algorithmic structure rests on three fundamental presuppositions, with which we shall have further dealings:

> (1) *substrate neutrality*: . . . The power of the procedure is due to its *logical* structure, not the causal powers of the materials used in the instantiation, just so long as those causal powers permit the prescribed steps to be followed exactly;
> (2) *underlying mindlessness*: Although the overall design of the procedure may be brilliant, or yield brilliant results, each constituent step, as well as the transition between steps, is utterly simple . . . ;
> (3) *guaranteed results*: Whatever it is that an algorithm does, it always does it, if it is executed without misstep. An algorithm is a foolproof recipe. (50–51)

Dennett argues that although the multiplicity of processes that constitute natural selection are algorithmic, that is, abstract, logical, formal, mindless systems, programs, which have determinable principles or even "laws" to direct them, they have no material substrate, that is, no particular reliance on any particular material form, and no determinable result. Dennett is introducing a new, more formal understanding of causality, a causality that is identified with a program or directives rather than with the force of matter, a probabilistic formulation that will guarantee, not particular results, but general tendencies over large numbers, given enough time. One must agree with Dennett that there is something fundamentally mindless and automatic about the Darwinian system, and that this is one of its explanatory advantages: it explains the individual and the life form entirely in terms of the existence of preindividual forces and processes. Dennett is also quite correct to recognize that the mindlessness of these processes renders no category, including the most hallowed of philosophy, untouched. All reason, conscience, nobility, all the human virtues and inventions are the long-term effects of the same kind of automatic mindlessness that regulates the existence of the most humble bacteria. What is "dangerous" about Darwinism, that is, what is in danger of being repudiated or repressed in Darwinism, is that it sets the whole of cosmology into a framework of forces that are incapable of being controlled by its participants:

> Darwin's idea had been born as an answer to questions in biology, but it threatened to leak out, offering answers—welcome or not—to questions in cosmology (going in one direction) and psychology (going in the other direction). If *re*design could be a mindless algorithmic process of evolution, why couldn't that whole process itself be the product of evolution, and so forth, *all the way down?* And if mindless evolution could account for the breathtakingly clever artifacts of the biosphere, how could the products of our own "real" minds be exempt from an evolutionary explanation? Darwin's idea thus also threatened to spread *all the way up*, dissolving the illusion of our own authorship, our own divine spark of creativity and understanding. (Dennett 1996, 63)

I agree with Dennett that if the forces of natural selection function as Darwin outlined, then mind itself must be the result of the same relentless processes of weeding, extermination, recombination, and variation that Darwin attributes to bodily existence. Yet Dennett's reductionism—for in many ways he is one of the more articulate exponents for the reduction of biology to genetics and of natural selection to algorithmic formulation—is not without its own problems. For all its commitment to prepersonal, subliving forces, a position, ironically, he shares (and acknowledges that he shares) with Nietzsche, he cannot provide a line of separation between processes, which, as I shall argue in chapter 8, are not always or usually reducible to step-by-step stages without a loss of cohesion and continuity and without losing a directionality that may be erased by decomposing analogue continuity into the digitized, discrete, atomized steps.

Dennett himself asks how it is that algorithms can be distinguished from processes: "Won't *any* process be an algorithm?" "Are there any limits at all on what may be considered an algorithmic process?" To which he answers: "No; if you wanted to, you could treat any process at the abstract level as an algorithmic process. So what? Only some processes yield interesting results when you do treat them as algorithms, but we don't have to try to define 'algorithm' in such a way as to include only the *interesting* ones . . . The problem will take care of itself, since nobody will waste time examining the algorithms that aren't interesting for one reason or another. It all depends on what needs explaining" (1996, 57, 59).

Dennett is, I believe, fundamentally correct about the essential mindlessness and directionlessness of the processes of both individual variation and natural selection, but it is not at all clear that his reduction of this mindlessness to a calculable set of abstract procedures, like Smith's rational free market, is without drastic cost. Dennett suggests that any process, or even any product, *can* be explained by algorithmic procedures, and that common sense or reason will dictate what will be the interesting results of such a reduction and distinguish them from the less significant results. But this implies that the algorithmic reduction occurs without cost, without residue, through a certain self-evidence about its relevance and value. As I discuss in further detail in chapter 8, the reduction of a continuity to step-by-step halts necessarily elides the dimension of duration, which is also a reduction of the active power Darwin attributes to the dynamic, forward movement of time. Dennett's algorithms, and indeed the algorithmic bases of computer-generated life simulations, though they may affirm a temporal order (one step after another), nevertheless deny the dynamic and continuous nature of temporal change. After all, they are programs that would run in the same way whenever they are initiated, even if they generate different results. The reduction of the temporal dynamism of matter, in its organic and inorganic forms, to abstract procedure necessarily loses sight of the concrete specificity of the object of Darwin's writings.[10]

This reductionism leads to a curious if unacknowledged recommitment to the very dualism Dennett seeks to problematize in those humanist readings of Darwinism (like his self-styled adversary, Stephen Jay Gould), to which he so strongly objects. It was this idea, that man's intellect as well as his body was formed by the chance-like operations of natural selection, that Darwin's colleague and the "cofounder" of evolutionary discourse, Alfred Russel Wallace, found unpalatable in his work.[11] Wallace (1889, 31) claimed, along with most theologians, and indeed most contemporary liberals and humanists, that mind itself must be the product of something higher than the forces of evolution: man's "intellectual and moral faculties . . . must have had another origin . . . in the unseen universe of Spirit."[12] Up to a certain point of organization or complexity, most scientists or philosophers are prepared to agree that teleology and purposiveness have no place in biological explanation. But it is significant, as Dennett makes clear, that many philosophers and scientists cannot accept the same of mind. Many spend considerable energy claiming that man's intelligence, or his social and cultural capacities, thrust him out of the evolutionary chain and into a mode of self-propelled evolution. Indeed, this seems to be the view of many of the current generation of social Darwinists, including the most well-known: Stephen Jay Gould, Edward O. Wilson, and Richard Dawkins. For example, Gould (1991, 63) states, "I am convinced that comparisons between biological evolution and human cultural or technological change have done vastly more harm than good—and examples abound of this most common of intellectual traps . . . Biological evolution is powered by natural selection, cultural evolution by a different set of principles that I understand but dimly."

It is ironic that while Dennett provides one of the more rigorous philosophical readings of Darwinism and has acknowledged and explored the "danger" of Darwin's idea, the threat it poses not only to received religions but also to those humanists who wish to attribute a post- or nonevolutionary status to the products of mind or reason, this was, for Dennett, the limit of Wallace's version of evolution: Dennett himself exempted mind from the operations of evolution. He submits mind to the same exigency when he distinguishes the biological evolution of species from what he describes, following Richard Dawkins's (1989) usage, as the "memetic" evolution of cultural and mental concepts. Dennett effectively reproduces precisely the mind-body split that he so convincingly criticizes in Wallace, Gould, and other evolutionary thinkers. He argues that the evolution of concepts is subject to the same principles of evolution as the evolution of biological entities. With this claim I have no disagreement. However, he seems to present the evolution of ideas in a landscape separate from the evolution of biological beings, when in fact, the evolution of concepts and cultural activities is simply the latest spiral or torsion in the function of one and the same biological evolution. Memes, atomistic conceptual units, ideas such as justice, democracy, science, and so on, are considered analogues of genes, rather than the ramifying products of genes. Memes are to mind what genes are to bodies! They retain, in Dennett's formulation as in Dawkins's, an uncanny resemblance to Platonic ideas:

> Meme evolution is not just analogous to biological or genic evolution, according to Dawkins. It is not just a process that can be metaphorically described in these evolutionary idioms, but a phenomenon that obeys the laws of natural selection quite exactly. The theory of evolution by natural

> selection is neutral, he suggests, regarding the differences between memes and genes; these are just different kinds of replicators evolving at different rates. And just as the genes for animals could not come into existence on this planet until the evolution of plants paved the way . . . so the evolution of memes could not get started until the evolution of animals had paved the way by creating a species—*Homo sapiens*—with brains that could provide shelter, and habits of communication that could provide transmission media, for memes. (Dennett 1996, 345)

Memes are housed, transmitted, and reproduced in brains; brains are, as it were, their vehicles of movement. But what relation do they owe to the specificity of those brains? How substrate-neutral can ideas be? The question remains: Is cultural evolution linked to biological evolution? Are they one and the same system, as Dennett implies but does not explain? Two different systems that have two environments and two modes of "natural selection," as Gould suggests? Is cultural evolution independent of biological evolution, something able to be controlled, as biology has thus far not been? Have we left behind the functioning of natural selection with the advent of technologies that have refined and controlled nature, including our own biologies? Does the evolution of mind follow a path of development different from the evolution of the body? Has mental evolution superseded, speeded up, transformed, protected, biological evolution? Or is mind, intelligence, part of the body itself, part of behavior, adaptedness, and adaptability?

The Descent of the Human

Darwin's response to the question of mind and of mental evolution is more subtle and complex than Dennett acknowledges. While Darwin devotes much of *The Descent of Man* to addressing the development of man's bodily morphology from primate ancestors, a good deal of his reflections are devoted to exploring the evolution of man's moral and intellectual characteristics, precisely those elements most dramatically at stake in evolutionary psychology, a contemporary heir of social Darwinism. But instead of positing a separate behavioral, psychological, intentional arena in which biology plays out its latest human incarnations, Darwin himself sees the moral and the intellectual as integrally bound up with the body's evolutionary development, contingent on the development of language. Mind is itself a function of the ever-growing elaboration of the body rather than a separate or self-contained system of its own.

Darwin's stated intent in *The Descent* is to "see how far the general conclusions arrived at in my former works, were applicable to man" (DM 1:3), that is, to see to what extent man, too, must be considered part of the world of animals and of nature from which he had been severed by eighteenth-century philosophy, particularly the philosophy of the Enlightenment. *Origin*, while carefully refraining from any mention of mankind, nevertheless established a panoramic horizon of examples, a vast interlocking of species, varieties, plants, and animals, whose only missing component was man. His purpose in *The Descent* is to rectify this self-conscious omission: "The sole object of this work is to consider, firstly, whether man, like every other species, is descended from some pre-existing form; secondly, the manner of his development; and thirdly, the value of the differences between the so-called races of man" (2–3).

Man's relation to the natural world is one of containment (man is in a dynamic and ever-changing natural milieu) and one of degree (man's resemblances to and affinities with the natural world are greater than his divergences from it): "In every single visible character man differs less from the higher apes than these do from the lower members of the same order of Primates" (DM 1:3), a claim that has genetic confirmation in the recent decoding of the human genome. Moreover, as Darwin makes clear at the very beginning of his text, an exploration of the descent of man would necessarily have, as part of its project, an analysis of the development of sexual selection and the manifest and latent differences between males and females and, through this concept, a quite profound and complex analysis of the development of racial differences. Although there is much that is problematic and historically dated in his analysis of racial differences, it appears that Darwin's work is remarkably open to the ways race tends to be considered today, in terms of racial differences, much more so than various social Darwinist and eugenicist reductions, which seek to define relations of superiority and inferiority in highly specious evolutionary terms. I will reserve a more detailed discussion of the question of racial difference and its connection to sexual selection—in other words, the ways Darwin's writings link together sexual with racial difference—for the next chapter.

Darwin's most basic argument is that man has been distinguished from animals and from the natural world by a number of different criteria, according to different philosophical and scientific positions, but these various distinctions are a matter of degree rather than a fundamental difference in kind, and are thus plausible effects of the operations of natural selection. He begins with an analysis of the bodily morphology of man. Not only are there remarkable resemblances in the embryonic and developmental structure of humans and animals, particularly mammals, but there is also a remarkable similarity in the functioning of key organs. There are even remnants—residual or rudimentary organs—often in an atrophied or undeveloped state, that testify to some other, historically prior form. (The history of teratology abounds with examples of rudimentary or supernumerary organs, e.g., the undeveloped organs of the opposite sex, rudimentary horns or tails, extra nipples in men or in women, body hair, even the appendix—indeed, as Darwin suggests, the whole external ear.)[13] His explanation for these homologies and for the existence of these rudimentary structures is that man is descended, not from apes themselves, but from some common progenitor of man and the ape, and that mammals and mankind, though sharing common ancestors, developed in quite different ways, facing quite different pressures from natural selection, and thus varying from their ancestors by ever greater degrees. Occasionally, there may be a regression back to a prior form, the emergence of a living fossil (we now have a genetic explanation for this regression), but this is rare.

But his most convincing arguments about man address, not man's bodily morphology, but his intellectual and moral capacities, to which they are necessarily linked. He devotes chapters 2 and 3 of *The Descent* to these arguments, where he claims that "there is no fundamental difference between man and the higher mammals in their mental faculties" (DM 1:35), and by "fundamental" he means unbridgeable, unobtainable by small gradations, gradual increments, or elaborations. There is only a difference in degree, not in kind, between the mental and moral capacities of man and those of animals. Darwin wants to correlate the development of higher mental functions with the increasing elaboration of the brain and its capacity to be dissociated from automatic, that is, instinctive behavior: "Little is known about the function of the brain, but we can perceive that as the intellectual powers become highly developed, the various parts of the brain become connected by the most intricate channels of intercommunication; and as a consequence each separate part would perhaps tend to become less well fitted to answer in a definite and uniform, that is instinctive, manner to particular sensations or associations" (38).

The brain's capacity to generalize, its removal from the circuit of automatic or reflex response, provides the basis for its capacity for generating other, more complex and less predictable, more innovative and problem-generated behavior. So Darwin implies that with the gradual enlargement and increasing "intercommunication" of parts of the animal brain, the higher emotions, and reason itself, gradually emerge. The brain's evolutionary elaboration implies that it moves further and further from automatic or instinctive behavior to generate innovative and unpredictable reactions.[14] Animals of various kinds, from both higher and lower orders, exhibit either incipient or full-blown emotions and the qualities that precondition intelligence, such as maternal affection for offspring (loyalty, kindness, care), love and the desire to be loved, wonder, curiosity, guilt, shame, embarrassment, imitation, attention, memory, and imagination; these are all emotions that man shares in common with other living creatures.[15]

Rather than vaunt reason as a set of abstract, logical principles available to man alone, Darwin is perfectly prepared to suggest that there are elements or degrees of reason in much of animal behavior, from learning to problem solving, from the use of tools to the capacity to communicate. Yet, it is believed that there are major factors that definitively separate the animal from the human: "It has been asserted that man alone is capable of progressive improvement; that he alone makes use of tools or fire, domesticates other animals, possesses property, or employs language; that no other animal is self-conscious, comprehends itself, has the power of abstraction, or possesses general ideas; that man alone has a sense of beauty, is liable to caprice, has the feeling of gratitude, mystery etc; believes in God, or is endowed with a conscience" (*DM* 1:49).

Out of all these characteristics that some claim distinguish man from animals, contemporary sensibility tends to single out the capacity for language as the defining difference, which makes possible abstract reasoning and cultural communication and most directly facilitates learning. It is primarily the complexity, richness, abstractness, and precision of language that generates the remarkable reflective and cultural capacities of mind in man. Darwin's approach here is ingenious: he wants to suggest that it is quite clear that some of the rudiments or preconditions of language, and all the other contenders for distinctively human attributes, are there in animals, simply awaiting adequate stimulus and development. Animals can certainly be considered to communicate, to express and to emote using sounds (dogs and cats, for example, have learned to articulate only since domestication). Moreover, in animals, many signs or elements of their systems of communication are not instinctive but learned. Darwin provides the evidence of songbirds, in whom the power to sing may be regarded as instinctive, but whose distinctive melodies are clearly learned and can even be learned by and transmitted from one species to another (*DM* 1:55). Darwin does not contest that it is only mankind who develops "articulate language," language that Roman Jakobson (Jakobson and Halle 1956) understood in terms of double articulation, the capacity to select and combine from the most elementary phonemic level to the most developed level of phrases and sentences. He argues, however, that this capacity has been developed gradually out of the vocal and auditory raw materials provided in the animal kingdom: "As monkeys certainly understand much that is said to them by man, and as in a state of nature they utter signal-cries of danger to their fellows, it does not appear altogether incredible, that some unusually wise ape-like animal should have thought of imitating the growl of a beast of prey, so as to indicate to his fellow monkeys the nature of the expected danger. And this would have been a first step in the formation of a language" (DM 1:57).

There is a *double* evolutionary structure at play in the existing forms of language. Language is made possible through the transformation, the adaptation, of the vocal and auditory organs of higher-order animals, coupled with the growing intelligence of the brain, so that language and what we might call cognitive structures emerge, as it were, simultaneously or in close cooperation, and heighten themselves through a productive spiral. The convenience of linguistic forms for the enhancement and complexity of representations and concepts, and the feedback structure in which intellectual demands find their precise articulation, rendering them more subtle and complex, generate an escalating emergence of what had not existed before: languages in all their complexity, conceptual schemas of increasing difficulty, abstract schemas for intentional action. This attests to the second twist in the evolutionary structure: that once a primitive language emerged from man's earliest progenitors, this language, and its descendants, were themselves subjected to natural selection, to development, use, supersession by competing languages, by transformation, development, extinction, or colonization.

The line we draw to divide language from sounds, and life from matter, is relatively arbitrary. In both cases, once this line exists, the operations of natural selection provide criteria for their fitness, their adaptation to socioenvironmental needs and their power and range relative to competing terms. Indeed, it is natural selection itself that draws the line, that ensures that the transformation of sound into language or matter into cellular life has advantages over mere sounds and unstructured, nonorganic material forces. If language is a defining characteristic of the human, then Darwin has shown that language itself can be a gradually accumulated acquisition, slowly developing in complexity and structure, in accordance with developing intellectual capacities and escalating social requirements.

If language does not serve to definitively distinguish the human from the nonhuman animal, then, many of Darwin's contemporaries argued, it is man's moral capacities—for conscience, sorrow, regret, guilt, and shame—that serve to distinguish the human and to elevate it above the animal. But Darwin's strategy is the same here as elsewhere: whatever characteristic we may regard as a defining one, a unique quality or attribute, he suggests, may be found in a less developed form elsewhere in the animal kingdom. His strategy is always to transform a difference in kind into a difference of degree. He develops the same arguments regarding man's moral capacities as he does regarding man's linguistic or biological capacities: all the various moral senses we may consider to be unique to man are already prefigured in animals. These qualities differ only in degree: "The following proposition seems to me in a high degree probable—namely, that any animal whatever, endowed with well-marked social instincts, would inevitably acquire a moral sense or conscience, as soon as its intellectual powers have become as well developed, or nearly as well developed as in man" (DM 1:71–72). The genealogy of this "moral sense or conscience" in man requires only a few assumptions:

1. That social animals have a kind of sympathy with each other.
2. That, as they become more mentally developed, they come to understand that long-standing social bonds are perhaps more valuable than the momentary satisfaction of individual desires.
3. That, with the acquisition of language, a notion of the common good would develop.
4. That, from this notion, habits of moral sense or conscience would develop.

If these conditions are met, Darwin implies, there could have been, or could be in the future, a different path of development to morality and conscience, the very possibility of, say, an insect ethics. This is perhaps one of the most significant sentences in *Descent*: "It may be well first to premise that I do not wish to maintain that any strictly social animal, if its intellectual faculties were to become as active and as highly developed as in man, would acquire exactly the same moral sense as ours. In the same manner as various animals have some sense of beauty, though they might have a sense of right and wrong, though led by it to follow widely different lines of conduct" (DM 1:73).

In other words, there is nothing particularly God-given or rational about human morality. It is a form, and by no means the only form, that can be called moral. Just as animals have widely different interests, tastes, and needs, so too the development of a different path to sociality and to intelligence may have developed an entirely different form and direction had they occurred in, say, a nonmammalian direction or, within mammals, according to the evolutionary development of, say, the rabbit or dog rather than the human. And no doubt, by implication, those moral forms that reign as "natural" or self-evident today are themselves in the process of evolution in line with human evolution. Morality, like reason and like other higher faculties in mankind, must be seen as resources that aid or hinder group and individual survival. As Darwin makes clear, morality and intelligence themselves have value only insofar as they play a pragmatic role in the existence of individuals and groups. It is significant that if a survey of comparative moralities were available, Darwin suggests, those acts, impulses, and desires that strengthen a culture's perception of its common good would become good and moral. In the past, and moreover, in the present, this has justified the most horrendous brutalities as the enactment of a moral good. He cites slavery and the torture of one's enemies as near universal impulses: the good is practiced with one's relatives, friends, and social group; outside this group, all manner of immoral acts are not only possible but actively encouraged. This provides a strange anticipation of Nietzsche's own genealogy of morals:

> The difference in mind between man and the higher animals, great as it is, is certainly one of degree and not of kind. We have seen that the senses and intuitions, the various emotions and faculties such as love, memory, attention, curiousity, imitation, reason etc, of which man boasts, may be found in an incipient, or even sometimes in a well-developed condition, in the lower animals. They are also capable of some inherited improvement, as we see in the domestic dog compared with the wolf or jackal. If it be maintained that certain powers, such as self-consciousness, abstraction etc, are peculiar to man, it may well be that these are the incidental results of other highly-advanced intellectual faculties; and these again are mainly the result of the continued use of a highly developed language. (DM 1:105)

The human is fundamentally an elaborate and relatively adaptable animal, an animal whose future is not within its own control, though cultural systems cannot function without a certain level of (voluntary) social participation. This participation, though, exerts less control over human social arrangements than the self-organizing structures of human "products": language, the economy, law, justice, and so on. These are not the emergence of "higher" qualities but the elaboration and expansion of certain animal impulses. And just as the human is an elaboration, a becoming-other of animal impulses to social and moral behavior, so too the human is in the process of becoming other-than-human, of overcoming itself. In the following chapter, I turn in more detail to the sexual and racial divisions of the human, and in part 2, to more discussion of this beyond-the-human.

3. THE EVOLUTION OF SEX AND RACE

Taken together, the principles of individual variation, heritability, and natural selection provide an explanation of a series of processes and interactions that are fundamentally mindless and automatic, and yet are also entirely unpredictable and inexplicable in causal terms or in any terms that atomize or isolate units, steps, or stages. Darwin conceptualized a machinery of natural forces—no longer gravity or mechanics, no longer precisely predictable—that, when they operate as a complex, as an assemblage, produce both massive variation and the beauty and elegance of life adapted in its most intimate contours and features to its environment. Natural selection does not simply limit life, cull it, remove its unsuccessful variations; it provokes life, inciting the living to transform themselves, to become something other than what they once were, to differentiate themselves by what they will become. Natural selection is not simply a passive background or context within which individual variation unfolds, a mere landscape that highlights and positions the living being, setting it into competition for scarce resources; it is an active dynamism that sets goals and provides materials and incentives for the ever-inventive functioning of species. Natural selection entails that the material world, and the other organisms by which a being is surrounded, function as a positive provocation to the self-overcoming that is the most basic characteristic of life, this self-overcoming attesting to the irreducible investment of life in the movement of time, its enmeshment and organization according to the forward direction of time.

Sexual Selection

Natural selection is rendered more intricate and complicated through the input of its two particular variants: artificial selection or domestic breeding and sexual selection. It is not accidental that Darwin begins his account of the origin of species with an elaborate discussion of domestic breeding. Artificial selection is the selective breeding of life forms through the human introduction of the specific criteria for selection, whether in terms of beauty, adaptedness to man's needs, or any criteria human breeders may choose. Darwin begins here because artificial selection seems to provide a more simple and well-documented model for understanding the more generalized, overarching, slower, and less visible relations of natural selection.[1] Instead of being construed as polar opposites, as cultural and natural binaries, as would be the case today in the near ubiquitous division of the study of culture from the study of nature, natural and artificial selection are regarded as two versions of the same thing, the artificial functioning according to the same principles as the natural, modifying only the criteria for selection. Underlying the utilization of manifestly cultural activities for explaining natural ones is Darwin's belief that, whether consciously or not, culture, and cultural production, manifested most directly in the history of languages or in the operation of economies, must submit itself to the same temporal exigencies as nature itself. It is not that one can read culture directly from nature, for nature does not incline itself to any particular kind of culture; it is simply that cultural relations, especially cultural systems (the economy, language, motor or pedestrian traffic, etc.) obey the same imperatives and forces as organic and temporally sensitive inorganic systems (natural populations, ecological systems, meteorological systems, etc.).

We exercise a form of artificial selection whenever we breed living beings for our purposes, whenever we select them for particular characteristics we need, like, or prefer (OS 49). Darwin suggests that the capacity for artificial selection exists only because there is already in natural breeds both the tendency to individual variability and the tendency to inherit the results of such variation. In artificial selection, we select those characteristics that appeal to us, and, through careful breeding with similar individuals, we can produce crops and animals that suit, perhaps while transforming, our needs and wants. But in doing so, we unconsciously borrow from, and add a complication, or rather, other criteria, to the operations of natural selection. We must nevertheless submit our intentions for specific outcomes to the indirect exigencies of the evolutionary process.

Artificial selection is largely based on the selection of visible, phenotypic characteristics, characteristics confirmable by observation—thickness of coat, number of fruits, the brightness of flowers—rather than any internal or invisible, genetic capacity to survive natural pressures. The plants and animals selected need not be the fittest in terms of their capacity for long-term survival or the production of the most surviving progeny, but they may nevertheless survive into the future just as prolifically through the privileged and focused breeding capacities human intervention has engineered. This remains as true today as in Darwin's time. Agricultural breeding is interested in the artificial or selective breeding of animals based only on measurable results: bigger bulk, faster growth, more protein, qualities that, if not visible to the naked eye, are certainly measurable using instruments: "On the view here given of the important part which selection by man has played, it becomes at once obvious, how it is that our domestic races show adaptation in their structure or habits to man's wants or fancies . . . Man can hardly select, or only with much difficulty, any deviation of structure excepting such as is externally visible; and indeed he rarely cares for what is internal. He can never act by selection except on variations which are first given to him in some slight degree by nature" (*OS* 58).

If domesticated breeding provides a commonly occurring scientific laboratory for the observation and manipulation, though probably not the production, of variation, a relatively controlled space for experiments in selection, in other words, if it can be regarded as a subbranch or "species" of natural selection, then sexual selection functions both as another inflection and complication of natural selection, but also, at times, as a principle compromising natural selection. If artificial selection augments and mediates natural selection through the production of human selective criteria, sexual selection has a more complex and indirect relation to natural selection. Sexual selection may function in opposition to natural selection: it may privilege members of either sex who may not be the fittest in terms of strength, health, or ability, but may in some other sense function as the more attractive. This other force at work in the life of species sometimes complements and extends natural selection, but at other times problematizes it. It forces us to shift or reevaluate the meaning of central concepts within the theory of natural selection—primarily fitness, struggle, and selection itself. Sexual selection adds an individual, idiosyncratic element to the operations of natural selection and, moreover, provides a different set of criteria for what counts as success or fitness in the evolutionary schema.[2]

Sexual selection introduces a further modality of divergence in organic life that will have its ultimately incalculable effects on epistemology as much as on ontology. Sexual selection ensures that sexual difference remains, at least into the foreseeable future of the human, and beyond—irreducible and impossible to generalize into a neutral or inclusive humanity. Sexual difference entails that, from the "moment" there is the human—and even long before—the human exists in only two nonreducible forms. Two forms, then, which have their own interests, needs, organic body parts, and ways of negotiating the world through them, two forms whose interests cannot be assumed to be the same but may negotiate a common interest in collective survival. Two forms, which not only divide most of life into divergent categories, but also produce two types of bodily relation with the world, and two types (at least) of knowing. Ironically, it is on this point, but perhaps only on this one, that feminist work on sexual difference converges with that of the strongly antifeminist writings of E. O. Wilson, Dawkins, and others, who essentialize the characteristics of the masculine and the feminine in terms of a reproductive telos of sperm and ova. There is an irreducible difference between the sexes, and this difference is not only irreducible to one of its terms, in the case of sociobiology, its reproductive cells; it is also irreducible to any other level, whether cellular, morphological, cultural, or historical. Sexual difference is irreducible difference, yet it is not a measurable, definable difference between given entities with their own characteristics but an incalculable difference that reveals itself only through its temporal elaborations. It is that difference which, in the future, will have been expressed, will have articulated itself, but which, in the present, has only represented itself from the point of view of one sex.[3]

Sexual selection differentiates all species touched by its trace with an irreducible binarism that itself generates endless variety on either side of its bifurcation, and indeed produces variations—the intersexes—that lie between bifurcated categories. Moreover, as Darwin suggests, evolution never reverses itself: it never goes from more to less developed, from more differentiated to less differentiated. Sexual difference, once introduced as a modality of biological survival, is unlikely to be removed, only complexified, elaborated, developed further, perhaps even beyond the human. The posthuman future is more likely to be sexually differentiated (in whatever form) than anything else we recognize in the present. The Darwinian model of sexual selection comes to a strange anticipation of the resonances of sexual difference in the terms of contemporary feminist theory! It provides the outline of a nonessentialist understanding of the (historical) necessity of sexual dimorphism, the earliest intimations of Irigaray's understanding of the ineliminable variation of sexual difference, and its productive inventiveness for future forms of life.

Darwin introduces the notion of sexual selection in part to explain what he could not ascribe to natural selection: the emergence and stabilization of particular trends that have no real survival value per se. Sexual selection "depends on the advantage which certain individuals have over others of the same sex and species, in exclusive relation to reproduction" (DM 2:256). Not all differences between the sexes are the result of sexual selection; indeed, a large part of sexual bifurcation is the consequence of natural selection and the evolutionary advantages that sexual difference bestows on hermaphroditic or self-fertilizing modes of reproduction. The increase in bodily motility, speed, and strength may serve to enhance reproductive prospects, but Darwin suggests that these evolutionary changes also have survival value independent of their value as sexual enhancements, and must thus be regarded as consequences of natural selection.[4]

The acquisition and refinement of secondary sexual characteristics, those that excite the interest of the opposite sex and function primarily for that purpose, are the results of sexual selection. Although it is the effect rather than the cause of sexual dimorphism, sexual selection nevertheless functions as dimorphism's elaboration and site of variation. The dividing line between sexual and natural selection is an ambiguous one: what was once developed for its adaptive value may become elaborated and transformed through sexual selection, and it is also feasible, though no doubt rarer, that what was once the object of sexual selection—for example, color, hair, facial features—may serve an adaptive function as well. As Darwin acknowledges, "In most cases it is scarcely possible to distinguish between the effects of natural selection and sexual selection" (DM 2:257). Sexual selection is most manifest when the nonreproductive differences between members of one sex—usually but not always males—are selected independent of their adaptive value:

> There are many other structures and instincts which must have been developed through sexual selection—such as the weapons of offence and the means of defence possessed by the males for fighting with and driving away their rivals—their courage and pugnacity—their ornaments of many kinds—their organs for producing vocal or instrumental music—and their glands for emitting odours; most of these latter structures serving only to allure or excite the female. That these characteristics are the result of sexual and not ordinary selection is clear, as unarmed, unornamented, or unat-

> tractive males would succeed equally well in the battle of life and in leaving a numerous progeny, if better endowed males are not present. We may infer that this would be the case, for the females, which are unarmed and unornamented, are able to survive and procreate their kind. (257–258)

Sexual selection is posited on the prior existence of sexual dimorphism. As early as *Origin*, Darwin suggested that there is commonly, though not universally, an evolutionary advantage to the interbreeding of pairs over forms of self-generated or hermaphroditic reproduction. He suggests that it is the combination of inherited material from two individuals that generates much greater variation, more difference, and gives new individuals an evolutionary edge. It is a mechanism that ensures exponentially increasing variability, that necessitates that the heritable structure from each individual is different from that of each of its parents while in some respects resembling them both. It produces a machinery for genetic variability, for inducing more difference, greater variability, than is possible in hermaphroditic beings, which, when they reproduce singly, produce genetic clones of themselves. Even in the case of hermaphrodites that are able to self-fertilize, there is usually a crossing or interchange between two hermaphroditic individuals. Hermaphrodites in fact tend to closely approximate the genetic requirements of sexual reproduction: "Various species are hermaphroditic, such as the land-mollusca and earth-worms; but these all pair. As yet I have not found a single terrestrial animal which can fertilise itself. This remarkable fact, which offers so strong a contrast with terrestrial plants, is intelligible on the view of an occasional cross being indispensable; for owing to the nature of the fertilising element there are no means, analogous to the action of insects and of the wind with plants, by which an occasional cross could be effected with terrestrial animals without the concurrence of two individuals" (OS 132).

Sexual bifurcation provides better resources with which natural selection can work because it induces more variation. However, in rare cases, such as single-celled organisms that reproduce autonomously, there is also an evolutionary or selective advantage, insofar as those organisms, which tend to be evolutionarily stable over long periods of time, simply reproduce a form that continues to be successful. These organisms have managed to retain an advantage from their reproductive stability (at the expense of greater development, specialization, and variation), while all others gain an advantage from the increased variability that sexual reproduction introduces into populations. Darwin implies that sexual divergence must have occurred very early in the evolutionary schema, as is attested to by the fact that the vast majority of plants and animals are sexually dimorphic. This means that sexual differentiation must have occurred before the intricate splitting of species into plants and animals and before there was much detailed differentiation between animal species. Sexual division, with its correspondingly different reproductive capacities and morphological variations, entails potentially ever more divergent morphological structures. (Since Darwin, we have learned, as a result of (partially) understanding the structure of the chromosome, that many characteristics that seem unlinked to sexually transmitted characteristics are nevertheless inherited by only one sex, because those genetic characteristics are carried by the X or the Y chromosome and because they take dominant or recessive forms.)

Even without the complications of contemporary genetics, Darwin claims that sexual selection takes on two forms or has two characteristic modes of operation: first, the tendency to differentiate the sexes from each other, particularly in secondary sexual characteristics; and second, the tendency of members of the same sex to become differentiated and selected, according to their relative strength, beauty, cunning, or skill: "[Sexual selection] depends, not on the struggle for existence in relation to other organic beings or to external conditions, but on a struggle between the individuals of one sex, generally the males, for the possession of the other sex. The result is not death to the unsuccessful competitor, but few or no offspring" (OS 117).

Sexual selection does not, as natural selection might imply, eliminate the weaker or less fit; on the contrary, it rewards the more beautiful and the more attractive, relatively independent of their fitness. Sexual selection amplifies natural selection. It also produces and exaggerates differences that have no particular advantage as far as survival or adaptation are concerned. They induce perceptible differences that please and attract: "Thus it is, as I believe, that when the males and females of any animal have the same general habits of life, but differ in structure, colour, or ornament, such differences have been mainly caused by sexual selection; that is, by individual males having had, in successive generations, some slight advantage over other males, in their weapons, means of defence, or charms, which have been transmitted to their male offspring alone" (OS 119).

These early remarks by Darwin about sexual selection are based on the assumption, largely as unquestioned in his time as in our own, that it is the males who compete with each other over the females, and it is the females who select among the successful males, even though there is now, and was then, evidence that this was by no means universal. (Much of contemporary feminist Darwinism is devoted to seeking out those animal species where cultural associations of femaleness and passivity are problematized.)[5] There has been a good deal of historical research done on the impact of early feminist political struggles on the biological discourses of the nineteenth century and on the pressures under which Darwin was put around the time of the publication of *The Descent of Man* to provide a scientific or quasi-scientific account of women's social inferiority, in other words, to intervene directly into the political struggles of his day around "the woman question." The standard, knee-jerk feminist reading of Darwin today is that he sometimes sounds suspiciously like an apologist for his own culture's masculine privilege:

> In contrast to accepting his theory of natural selection, many feminist scientists have critiqued Darwin's theory of sexual selection for its androcentric bias. The theory of sexual selection reflected and reinforced Victorian social norms regarding the sexes . . . Expanding considerably on the theory first presented in the *Origin of Species*, Darwin specified, in the *Descent of Man*, how the process functions and what roles males and females have in it . . . According to the theory, the males who triumph over their rivals will win the more desirable females and will have the most progeny, thereby perpetuating and increasing, over numerous generations those qualities that afforded them victory. (Rosser 1992, 57)

And further:

> For Darwin patriarchy was entrenched, and there was close conformity between his findings on sex differences and those of conventional opinion. The sex stereotyping of the *Descent* is best understood in terms of an appreciation that evolutionary biology could be employed to counter the undue ambitions of women. Darwin's long-standing sympathy for women's education was now subordinated to the hard fact of female inferiority. (Erskine 1995, 100)

This reading has been prevalent in feminist egalitarian writings since the earliest publication of Darwin's works. Feminist egalitarians have remained wary of the implications of Darwin's account for understanding the relative social, cultural, and economic position of men and women, and indeed, it is true that biological discourses have been, and continue to be, used to constrain women's possibilities of change. It is not an unjustified claim that biological discourses have served to rationalize and justify the worst of assumptions regarding the relations of superiority and inferiority between sexes and races. But it is not clear how much Darwin himself succumbed to these assumptions. Moreover, although his own position seemed to reflect the liberal tolerance for which John Stuart Mill and Harriet Taylor were known, it is not his personal views about sexual difference that are the most significant here. Rather, what remains crucial and relatively unrecognized by feminists and others in his writings is the reconfiguration of culture in light of the fundamental openness he attributes to the natural world. Culture—whether patriarchal, class-based, or racist—is no longer the extension and completion of nature, the coloring in of the contours provided by nature. Nature is open to any kind of culture, to any kind of "artificiality," for culture itself does not find pregiven biological resources, but makes them for its own needs, as does nature itself. Culture produces the nature it needs to justify itself, but nature is also that which resists by operating according to its own logic or procedures.[6] A reconfiguration of nature as dynamic, of matter as culturally productive, of time as a force of proliferation, is thus central to the ways feminism itself may be able to move beyond the politics of equalization to more actively embrace a politics affirmative of difference elaborated in the most dynamic forms of feminist theory today.

Nevertheless, feminist theory in the present remains suspicious of the concept of sexual selection without seeing that it actually affirms the centrality of an individual level of choice and a small arena of relative control in the functioning of the impersonal machinery of natural selection, something not computable in algorithmic form. Darwin's position on sexual selection is highly ambiguous. It does posit the active competition of males with each other and the relative passivity of females who select from among competing males. It also makes clear that life, very early in its evolutionary movement, saw in sexual division and differentiating one means, probably the most ingeniously simple, by which inventiveness, creation, and the new coincide with the elaboration of life.

Rather than simply see Darwin as an apologist for the conservative values of his day—a position that is difficult for us to maintain even on a superficial reading, given his self-evident liberal-mindedness and his differences from the religious doxa pervasive when he wrote—that is, rather than regard him as another in the unceasing history of the patriarchal oppression of women (which he undoubtedly is), he needs to be addressed, as Freud has been in the past three decades by feminist theorists, as a scientist/theorist whose aim, whether successful or not, was descriptive, not prescriptive, and whose goal was not to render female conformity to his scientific expectations, but to provide evidence of the range and scope of the behaviors of both sexes. So,

I am less interested in providing criticisms of Darwin's sexism, or in criticizing his position at all, than in seeing how he proposes that sexual difference works, and what value there might be in his account for our current understanding of social, political, and biological relations between the sexes. My project here is not the critical endeavor of seeking out errors, biases, or mistakes in Darwinism, but rather, to see what of Darwin, and the philosophical figures that follow him, may be of use to a feminist politics of transformation, which may find his conceptions of time and becoming helpful in rethinking concepts of nature and culture, of human and animal, mind and matter, outside their more conventional feminist frameworks.

In *Descent*, Darwin's position solidifies and renders more conservative the one he outlined in *Origin*. There he claims that sexual selection works to differentiate the sexes from each other, to place the males (generally, at least, but with notable exceptions) in a position of rivalry and competition with each other, and to place the females (generally, but with notable exceptions) in a position of choosing among competing males. In *Origin*, success is primarily understood as the capacity to continue living, to survive; in *Descent*, however, the criterion seems to shift to the amount or quantity of offspring (incidentally, the criterion of success fixated on by sociobiology—a criterion that, by definition, gives a natural evolutionary advantage a priori to males insofar as males are capable of generating many more offspring than females, though not necessarily in ensuring their ongoing survival). Darwin himself, however, is more subtle than his followers: for him, the criterion isn't clearly measured by reproductive outcomes (i.e., progeny), but perhaps by reproductive taste! The most successful or sexually attractive animals are not clearly those who produce the most offspring, but those who attain the right to choose their sexual partners. Successful animals gain sexual access to successful animals, those they find most attractive (which may or may not coincide with their evolutionary fitness), and the result or effect of this may be, but need not be, reproductive success, more numerous progeny. Sexually successful beings are successful in the sense that their tastes, choices, preferences are more likely to be satisfied: "Secondary sexual characteristics . . . depend on the will, choice, and rivalry of the individuals of either sex" (DM 2:258).

Where Darwin gives males of most species the active power to compete with each other, to battle for favors with or access to the females, he gives females the passive power to select, to exercise their powers of discernment or discrimination, to sort from among those available the sexual partners they find the most attractive. The males tend to develop powers of struggle or display, abilities that either help them compete against other males or in some way differentiate themselves, or to attract the attention and interest of females. These qualities—strength, skill, and beauty—may be more advantageous in terms of leaving quantities of progeny, or at least access to possible reproductive success, than other survival skills. But the mere conquest of other males does not guarantee reproductive success, for the ability to attract the female must follow: "In a multitude of cases the males which conquer other males, do not obtain possession of the females, independently of choice on the part of the latter. The courtship of animals is by no means so simple and short an affair as might be thought. The females are most excited by, or prefer pairing with, the more ornamental males, or those which are the best songsters, or play the best antics . . . Thus the more vigorous females, which are the first to breed, will have the choice of many males; and though they may not always select the strongest or the best armed, they will select those which are vigorous and well-armed, and in other respects the most attractive" (*DM* 2:262).

Males perform activities of conquest and display, more or less successfully relative to other males, which ideally, but not always, function to attract available females. But, as Mayr (1997, 77) notes, these performances of sexual attraction, "performances of seduction," quite often go astray from their own excesses, their blind undirectedness: "It is easy to observe that males tend to display not only to [females of their own species], but also to females of related species, to males of their own or related species, and, in the absence of appropriate display partners, to even less appropriate objects. If females were lacking discrimination to a similar extent, an enormous amount of hybridization would take place."

Counterposed with the male activities of aggressive competition with other males, including battles of conquest as well as more symbolic rituals, and the male's capacity in many species for virile, colorful, erotic display are those activities that are more commonly assumed to be female proclivities, generally conceived in terms of the female's relative passivity, which leads her to assert her sexual preferences primarily by selecting from among eligible males for sexual partners: "The female, though comparatively passive, generally exerts some choice and accepts one male in preference to others. Or she may accept . . . not the male which is the most attractive to her, but the one which is the least distasteful.[!] The exertion of some choice on the part of the female seems almost as general a law as the eagerness of the male" (*DM* 2:273).

Darwin is not reluctant to admit that in certain species, specifically in the case of many species of bird, this relation is reversed, and the females are more highly colored and ornamented, stronger, and more eager in their courtship of males (DM 2:276). He does imply that the differential marking of males and females in species where they are so differentiated may be the consequence of sexually differentiated criteria of beauty, though he is reluctant to admit that the operations of sexual selection must be considered a two-way or symmetrical relation. Indeed, this lack of symmetry in nature must have some significance for thinking about sexual politics and for the project of equalization: "It may be suggested that in some cases a double process of selection has been carried on; the males having selected the more attractive females, and the latter the more attractive males. This process, however, though it might lead to a modification of both sexes would not make the one sex different from the other, unless indeed their taste for the beautiful differed; but this is a supposition too improbable in the case of any animal, excepting perhaps man, to be worth considering" (277).

Sexual selection adds more aesthetic and immediately or directly individually motivating factors to the operations of natural selection; it deviates natural selection through the expression of the will, or desire, or pleasure of individuals. While conforming in the long run to the principles of natural selection, sexual selection nonetheless may exert a contrary force to the pure principle of survival. Sexual success (having many sexual encounters, generating many progeny) may not coincide with organic success (living beyond infancy to reach reproductive age). It is this very question of what counts as success, what counts as fitness, that still haunts Darwinism even today.

Darwin notes that many features of animal appearance and adornment, even those that may in some way render the being less able to survive, more noticeable to predators, less able to protect itself, and so on than its dowdier yet more fit counterparts, nevertheless have survival value. In the case of the spectacular plumage of the peacock relative to the plainness of the peahen, Darwin's explanation is that even if its plumage and adornment make the peacock more vulnerable to attack, the more magnificent its coloring, the more the attractiveness of the peacock to the peahen is enhanced. Although it is or may be disadvantaged in the stakes of natural survival, it is positively advantaged in the stakes of sexual selection.

Darwin's statements are all the more contentious as he approaches the structures of sexual difference in the human (and here, it is clear that his position is deeply implicated in the cultural values of his time and in the debates around the woman question common in his day). These claims are at the center of the feminist critique that continues to be directed to his work.[7] And indeed, his statements reflect less any scientific analysis of the two sexes than a series of common opinions, even stereotypes. A series of selected quotations will make this clear:

> With respect to difference of this nature between man and woman, it is probable that sexual selection has played a very important part. I am aware that some writers doubt whether there is any inherent difference; but this is at least probable from the analogy of the lower animals which present other secondary sexual characters . . . Woman seems to differ from man in mental disposition, chiefly in her greater tenderness and less selfishness; and this holds good even with savages . . . Man is the rival of other men; he delights in competition, and this leads to ambition which passes too quickly into selfishness. These latter qualities seem to be his natural and unfortunate birthright. It is generally admitted that with woman the powers of intuition, of rapid perception, and perhaps of imitation, are more strongly marked than in man; but some at least of these facilities are characteristic of the lower races, and therefore of a past and lower state of civilisation.
>
> The chief distinction in the intellectual powers of the two sexes is shewn by man attaining to a higher eminence, in whatever he takes up, than woman can attain—whether requiring deep thought, reason, or imagination, or merely the use of the senses and hands. (*DM* 2:326–327)

And:

> It is indeed, fortunate that the law of the equal transmission of characters to both sexes has commonly prevailed throughout the whole class of mammals; otherwise it is probable that man would have become as superior in mental endowment to women, as the peacock is in ornamental plumage to the peahen. (329)

Although we cannot afford to ignore these statements, we should not be surprised to see these elements of his own culture seeping into these remarks, and indeed into much of his biological account. And though some of his comments, for example, about the education of women, are highly provocative, it is nevertheless possible for us to reread the question of sexual difference in quite different terms from those he provides. He seems to suffer the same dilemma as all the other proponents and adherents of sexual egalitarianism. Egalitarianism, the claim that men and women are, or are potentially, the same, always measures women in terms of how they conform to the characteristics and values of men. It is always an egalitarianism in which men's attributes and qualities are the norms by which women are defined as equal. There has yet to be an adequate evaluation of Darwin from the point of view of a feminism that questions the givenness of this norm and that insists on sexual difference rather than equality as the criterion of social and biological value. Darwin himself may well have opened up sexual difference not only to its morphological but also to its cultural irreducibility.

Darwin wants to link the question of sexual selection to two of the most momentous developments in human evolution: first, to the development of language; second, to the descent of the different races of man. To briefly outline the first claim in order to leave more time for the second: Darwin hints, without really developing it, that the facility of vocalization not only has survival value, but primarily functions as a sexual lure. He claims that it is likely that all animals in some way find music pleasurable. Sonorous rhythms and melodious articulations, as sexual attractors, may well form the earliest origins of audition, and pleasure itself may provide the primitive motivation for the development of language:

> The perception, if not the enjoyment, of musical cadences and of rhythm is probably common to all animals, and no doubt depends on the common physiological nature of their nervous systems. Even Crustaceans, which are not capable of producing any voluntary sounds, possess certain auditory hairs, which have been seen to vibrate when the proper musical notes are struck . . . With all those animals, namely insects, amphibians, and birds, the males of which during the season of courtship incessantly produce musical notes or mere rhythmical sounds, we must believe that the females are able to appreciate them, and are thus excited or charmed; otherwise the incessant efforts of the males and the complex structures often possessed exclusively by them would be useless. (DM 2:333)

Music, the pleasurable, rhythmical repetition of sounds, the variability of melody and cadence, was selected, not simply, as Darwin suggests in *Origin*, because of its survival value, that is, because these vocalizations served as forms of warning for the group, but also, perhaps above all, because of their appeal to the opposite sex and especially because of the pleasure of male sounds to female ears: "A strong case can be made out, that the vocal organs were primarily used and perfected in relation to the propagation of the species" (DM 2:330). This musicality, the articulation of sounds, not simply for information, but for expression and seduction, is part of the genealogy of language, part of the reason, perhaps, that language itself is always playful, always seductive, always has the capacity to convey and induce pleasure and sexual responsiveness: "The diversity of sounds, both vocal and instrumental, made by the males of many species during the breeding-season and the diversity of the means for producing such sounds, are highly remarkable. We thus gain a high idea of their importance for sexual purposes, and are reminded of the same conclusion with respect to insects. It is not difficult to imagine the steps by which the notes of a bird, primarily used as a mere call or for some other purpose, might have been improved into a melodious love-song" (67).

Some of the articulations, courting sounds, and behavior that one sex uses to interest, attract, and seduce the other are also highly pleasurable to human subjects. This may explain, as Darwin suggests, the extravagant cost of songbirds (*DM* 2:53). But above all, it explains the human susceptibility to vocal attraction, which results in human musicality, both vocal and, derived from it, instrumental: "With man song is generally admitted to be the basis or origin of instrumental music. As neither the enjoyment nor the capacity for producing musical notes are faculties of the least direct use to man in reference to his ordinary habits of life, they must be ranked amongst the most mysterious with which he is endowed. They are present, though in a very rude and as it appears almost latent condition, in men of all races, even the most savage" (333).

While its origin and ongoing survival implies that music must have exerted a powerful force in sexual selection, and have been successively enhanced and strengthened in the process, Darwin also argues that musicality has become closely associated with expression, affect, and emotion. Music stirs both the strongest and the most subtle emotions and elicits the power to express and communicate intensity. He explains that this capacity to express and elicit emotions may have had to do with man's ancestry and with the residual function of vocalization to stir fear, desire, or jealousy to a heightened degree in man's primitive ancestors. Whatever its origin, music for humans retains its similarity with instinctive articulations and the excessive, nonutilitarian pleasures potential to all higher animals. This inherent playfulness and erotic charm makes its way into music considered more formally, and into musical elements that also eroticize language:[8] "Music affects every emotion, but does not by itself excite in us the more terrible emotions of horror, rage, &c. It awakens the gentler feelings of tenderness and love, which readily pass into devotion. It likewise stirs up in us the sensations of triumph and the glorious ardour for war. These powerful and mingled feelings may well give rise to the sense of sublimity" (*DM* 2:335).

Sexual selection may account for the development of a number of nonadaptive, excessive characteristics and capacities (excessive, that is, from the point of view of bare survival but necessary from the point of view of reproductive success; he cites, as another example, the female selection of the beard [*DM* 2:373]). But Darwin makes it clear that sexual bifurcation, particularly the sexual division of labor and value that marks all known human cultures, is not simply the result of either natural or sexual selection, but can be explained in terms of cultural, that is, political forces (read: man has artificially selected weakness, vulnerability, through sexual choice and educational and social practices): "Man is more powerful in body and mind than woman, and in the savage state he keeps her in a far more abject state of bondage than does the male of any other animal; therefore it is not surprising that he should gain the power of selection" (371).

The sexes diverged through processes of natural selection. Sexual selection intervenes into, but does not override, natural selection, whose only criterion is the capacity for survival. Sexual selection enhances natural selection, even directing natural selection through the detour of individual taste and discernment. The results of this selective process are the ever-growing divergence of the sexes not only in primary but also in secondary sexual characteristics; more particularly, Darwin suggests, the effects of sexual selection are most manifest in the long-term production of racial diversity.

Fitness and Survival?

For Darwin, the question of fitness, or survival—what it is that selection selects for—is rendered more ambiguous and complicated with the acknowledgment of the deviating relation that sexual selection imposes on natural selection. It is no longer clear whether fitness means the survival of the individual organism (which natural selection oversees) or its capacity to reproduce, which may or may not entail the survival of the individual (sexual selection). This is another way of asking whether it is the survival of the individual or the survival of the species that is at stake in natural selection. In more contemporary terms, the question becomes whether it is the organism or the gene that is being selected.

It was George Williams (1966) who first drew attention to the fact that there is sometimes a clash between the interests of the organism, of the species, and of the gene, the reproductive unit of differential replication. Williams claimed, contrary to Mayr, that it was possible to assign to each gene an independent fitness value, even if it only ever worked in cooperation with other genes, which enabled him and those following to claim that although genes may never function singly, they are nevertheless obvious minimal "units of selection." Williams's approach implied that, as the units of selection, those genetic elements that survive the shuffling, crossing processes of meiosis are what is selected, enter the next generation, and are thus proliferated beyond their fleeting incarnation in a particular body. Organisms become the passing, mortal bearers of the immortal or potentially immortal germ line first proposed by Weismann (1893). Weismann, in close anticipation of contemporary genetics, a mediator in the so-called Modern Synthesis that drew together the theory of natural selection and the theory of genetics, believed that natural selection works because germ cells, the only units directing replication, make somatic cells which make organisms selected according to the specific pressures of their competitive environments.[9] Which germ cells survive into the next generation depends on how well the organisms that bear them succeed in the competitive struggle for survival and reproductive success. The source of heritable variation, the germ cells, are only indirectly selected through the success of the individual that carries them. The germ cells direct the behavior of somatic cells, but the somatic cells are incapable of affecting the germ cells. This is the famous barrier Weismann proposed between the line of inheritance, the germ line, and modes of somatic development. The germ line remains unbroken; indeed, Weismann refers to "the immortality of the germ line," which, in his conception, never creates new germ cells but "can only grow, multiply, and be transmitted from one generation to another" (xiii). Adaptedness is the result of the changing proportion of germ cells that survive and reproduce themselves in populations. The battle for survival has been subtly moved from the organism to the cellular level, hereditary units competing with each other to gain a place in the continuity of the immortal germ line:

> The germ line represents the skeleton of the species on which individuals get attached as excrescences. Any changes that are the result of outside influences are merely temporary and disappear from evolution when the individual reaches its end. External events are no more than transient episodes that affect particular life-forms but not the species, which goes on regardless of changes to individuals. The germ line has thus, it is alleged in Weismann's account, guaranteed the reproduction of identical cells. Although germ cells vary from species to species, new structures required by evolution are produced not by individuals, but by the hereditary arrangements contained in the germ cells. So while natural selection *appears* to be operating on the aspects of an adult organism, in actuality it is working

> only on the predispositions lying concealed in the germ cell. The germ line, therefore, is outside the reach of any variation that takes place in individuals of the species. (Pearson 1999, 6–7)

Williams's argument concurs with that of William Hamilton (1964), that, from the gene's perspective, it matters little whether it is carried in one body or another as long as it can transmit itself. If, as they argue, the phenotype is largely irrelevant to the genotype's long-term survival except indirectly, as its temporary bearer or vehicle, then there may be an evolutionary explanation for the emergence of cooperation, social organization, and, one of their favorite topics, altruism.[10] In line with Adam Smith's notion of self-interest which these theorists assume, the theory of natural selection needs to be able to explain how it is that the cooperative, even the self-sacrificing, altruistic behavior observed in many species may have survival benefits for the group as a whole, even if individual organisms die in the process. Williams argues that because cooperative behavior is vulnerable to what he calls "cheaters," organisms that do not reciprocate and put their interest above the collective, there needs to be some other explanation for the manifest altruism of certain species, particularly insects. If genes are the primary units of selection, there may be an explanation for the emergence of collective and altruistic behavior: cooperation may not be the best strategy for individual survival and success, but it is a powerful strategy for a genotype that seeks to imprint as many future beings with its shared genetic structure as possible.

Hamilton argues that this structure becomes clear with relatives, that is, with genetically connected organisms, and in this case, the relationship can be quite precisely numerically calculated: one's siblings share literally half one's genes, one's cousins a sixteenth, and so on. In sacrificing oneself for one's close relatives, one's shared genes may have a greater chance of survival than through their own reproductive success. Cooperation is explained in terms of degrees of genetic relatedness. Altruism is, ironically, merely apparent: the gene functions primarily to extend itself, and if that involves cooperation, even self-sacrifice, then this is simply an indirect way of attaining its own self-interest. The gene is fundamentally, even in the case of altruism, in the language of Dawkins, selfish!

The theory that natural selection serves the gene's point of view culminates in the surprisingly powerful and infamous text by Richard Dawkins (1989), *The Selfish Gene*, which acknowledges the work of Hamilton and Williams as its predecessors. For Dawkins, organisms are primarily vehicles, protective shells, the resources that genetic materials require for their transmission. Genes must be understood to be selfish in two distinct but related senses: they are selfish because their only interest and goal is their successful transmission and reproduction, and they are selfish because, Dawkins believes, genes produce organisms more and more adapted precisely in order to enhance their own capacities for replication. Organisms are construed as complex forms required for generating replicating mechanisms. True to the heritage of Weismann, and in resonance with Platonic forms themselves, genes have an immortality, whereas organisms are only ever ephemeral. Like Wilson's (1980) bodies that serve as vehicles for DNA to reproduce itself, for Dawkins (1989, 20), organisms are "the survival machines" of genes. Natural selection no longer functions for the good of the species or for the good of the organism; it operates primarily for the good of the gene: "To whose interests is the actual 'decision-making' of natural selection most directly responsive? It is not controversial that conflicts between genes and bodies (between genes and the phenotypic expressions of the genotypes of which they are a proper part) can arise. Moreover, no one doubts that in general the body's claim to be considered the principal beneficiary lapses as soon as it has completed its procreational mission. Once the salmon have fought their way upstream and successfully spawned, they are dead meat. They literally fall apart, because *there is no evolutionary pressure* in favor of any of the design revisions that might prevent them from falling apart" (Dennett 1996, 329).

There may be a conflict of interest in some species between what is of survival benefit for the organism and what is of benefit to the genes that help construct that organism. As Samuel Butler (1887) asks, is the chicken the egg's way of producing another egg? Or, in the more contemporary language of E. O. Wilson (1980, 3), is the organism DNA's way of making more DNA? Does this mean that natural selection responds to and functions primarily in aid of the dissemination of genes? Is Wilson's position rather than Darwin's more accurate and appropriate? How are we to select between these alternatives? It is clear that the gene has a perspective or interest that cannot be assumed to be the same as the organism's, but it is also clear that the organism and the species each has its interests in the machinery of natural selection. Which is to be privileged? Though Wilson, Dawkins, Dennett, and their colleagues have argued that this perspective has great explanatory power, it is also true that these other perspectives, and no doubt many others (why wouldn't there be a "protein's perspective" or a "cell's perspective" as much as there is a gene's point of view?), also exist and are irreducible elements of natural selection.

The gene does have an interest in propagating itself, but the gene is at the mercy of the organism, which may or may not be reproductively successful. Though leaving no progeny, nonreproductive organisms, even in Darwin's writings, can nevertheless be explained in terms of the evolutionary advantages they offer, not to the gene pool, but to the social group, whether close kin or only broadly genetically connected organisms of the same or related species. This must be, in part, an evolutionary explanation for the ongoing production and survival value of homosexuality and its near ubiquity in all known cultures, that there is no evolutionary disadvantage to social collectives in which it occurs, and indeed there may be positive social advantages. Such advantages may have to do with the value societies bestow on some, though not all, of its members acting for the collective good (Wilson's postulate of games strategy, like all such models, cannot differentiate among the different values of particular strategies of particular organisms for the collective good. They must universalize the game players, who are regarded as interchangeable or undifferentiated, playing according to the same rules, playing the same game), the relative value of the increasing specialization of populations, the reservation of nonreproductive roles for some of its members. Wilson, Dawkins, and others, moreover, cannot really appreciate the nongenetically oriented strategies that nonreproductive groups add to the social collective, for the question of the rates of successful reproduction and the maximization of related genes is the only concern for such sociobiological perspectives, which thus reduce all strategies, reproductive and nonreproductive, to the one criterion of reproductive success.

Darwin acknowledges that the place of nonreproductive organisms is crucial to the plausibility of his theory of natural selection. It represents the case of "one special difficulty [facing the theory of heritable descent and natural selection], which at first appeared to me insuperable, and actually fatal to the whole theory" (OS 352). Nevertheless, there are still neuters, sterile organisms, nonreproducing members of populations, and pervasive homosexual couplings in much of the natural world. It is Darwin's belief that these can be explained, and the theory of evolution as he knows it can be kept intact, using the principles he has already outlined. He authorizes a shift from the survival value of the individual to that of the kin group, the "family," as he describes it, which is close indeed to many of Wilson's claims: "Selection has been applied to the family, and not to the individual, for the sake of gaining serviceable ends. Hence we may conclude that slight modification of structure or of instinct, correlated with the sterile condition of certain members of the community, have proved advantageous: consequently the fertile males and females have flourished, and transmitted to their fertile offspring a tendency to produce sterile members with the same modification" (354). Although here he is primarily discussing insects, which are known for producing large numbers of sterile workers (bees and ants come to mind), he also claims that this makes sense "on the same principle that the division of labour is useful to civilised man" (358). The continual production by males and females of nonreproducing offspring can be explained in terms quite other than gene maximization, that is, in terms of the spreading of shared genes. The increasing specialization of tasks in social communities (from the insect world to the human), the ways nonreproductive social beings help render more proficient the reproduction of others, and the production of social requirements needed for the creation of an environment in which others can reproduce have as much or greater evolutionary advantage than a species in which all members remain undifferentiated and compete with each other for reproductive success.

Racial Differences

More interesting than and as contentious as Darwin's connection of sexual selection to the acquisition and elaboration of secondary sexual characteristics and its ability to deviate the attributes chosen by natural selection through the pleasure principle of sexual attraction is his linkage of sexual selection to the differences between races. Sexual selection—taste, individual choice, preference, aesthetics—may have dictated that what were once slight variations in individual racial characteristics (color, features, proclivities) could provide criteria by which males and females choose each other as sexual and reproductive partners. Darwin suggests that the differences between human races cannot be explained according to God-given types, fixed categories, or any racial essence. As with the descent of every species, all the human races are the consequence or result of the selection of various degrees of difference, which are categorizable through the aggregation of resemblances, though not necessarily with great accuracy, as (relatively) distinct races.

Human racial differences, while they may be the result of the selective diversification and intensification of what were once individual differences, cannot be regarded as distinct species, subspecies, or even varieties. Recall that, for Darwin, what enables a more natural mode of classification of individual differences are two kinds of linkages: one historical, the other contemporary. Species, subspecies, and varieties can be linked through a common genealogy and ancestry, coming from related parental stock; they are classifiable together through the commonness bestowed on them by the possibility of fertile reproduction, by the exchange of genetic material. It is significant that although the human genome project has barely completed its first draft of the inventory of human genes, it has become clear that the genetic material circumscribing racial differences is only a minute fragment of the genome: humans of all races have much more in common with each other than the infinitesimally small genetic components that distinguish them. Individual differences, at least on a first draft reading of the genome, are more significant than racial or cultural differences. Racial differences are, in other words, entirely transformable, entirely open to historical and social transformations, though they must always be mediated by sexual relations.[11]

Darwin provides a striking, and commonly unrecognized, reading of the near universal representation of race as a representation of relations of inherent or obvious superiority and inferiority. Racial differences are themselves the long-term result of sexual rather than natural selection. In this sense, Darwin stands in stark contrast both to the racist and imperialist assumptions that his work was manipulated to confirm—assumptions regarding the evolutionary superiority of white races, the desirability of eugenic control of the reproductive possibilities of cultural undesirables, the future directions of racial development, and so on—and to more contemporary readings that, in diminishing the biological bases of racial differences, end up attributing them solely to the influences of culture and environment when there are clear biological factors relevant to the ways these cultural and environmental factors have their effects on different bodies. Sexual differences, the bifurcation of the human into male and female morphologies, are an ubiquitous element in all higher animals. The playing out of the preferences, ideals, and charms of sexual attraction induces the ever greater diversification of racial differences. Racial categories are the long-term effect of increasingly intensified and diversified sexually selective criteria. These criteria are generally not linked to mere survival, but are sexual.[12] They have less to do with material survival than with sexual preference and reproductive success.

Darwin provides strong arguments to show that racial differences cannot be attributed directly or solely to the selective pressures imposed by the environment. In this claim he is at odds with the conventional "evolutionary" explanation of the differences in bodily morphology in different races, that these characteristics have survival value in themselves:[13] "If . . . we look to the races of man, as distributed over the world, we must infer that their characteristic differences cannot be accounted for by the direct action of different conditions of life, even after exposure to them for an enormous period of time" (DM 1:246). External conditions, the direct effects of climate, geography, temperature, and so on, must be limited to negative selective forces rather than seen as positive inducements to change in particular directions (indeed, this was the basis of Darwin's critique of Lamarck, who claims that external forces create individually heritable changes). At most, changes in the environment induce variability, "and natural selection will then accumulate all profitable variations, however slight, until they become plainly developed and appreciated by us" (OS 173–174).

Racial differences cannot be seen as the direct result of the selective capacities of extremes of environment, that is, as the consequence of external, nonintrinsic factors. Darwin never accepted the claim, prevalent among his contemporaries and teachers, that the differences between different races are a measure of cultural, intellectual, and moral advancement. On his broad travels aboard the HMS *Beagle*, Darwin was struck by the diversity of practices and rituals, the variety of modes of intelligence, social regard, and cultural forms he encountered in his travels, and by the capacity, as he perceived it, that all races had for immense change and development.[14] Darwin believed in the plurality, or perhaps even the relativity, of social, moral, and aesthetic categories. It is hard to impose a notion of progress, or of superiority and inferiority, when the only criterion of success is the ingenuity of adaptation and the only necessary proof of adaptation is current existence.

Darwin suggests that racial differences may be the consequences of sexual selection more than of natural selection. It may be precisely the sexual appeal or attractiveness of what were once simply modalities of individual variation, however slight they may have been to begin with, that explains the historical variability and the genealogical emergence of racial differences. Racial differences may have been those differences that have been actively selected by individuals and perhaps amplified through geographical dispersion and subsequent isolation from our racially less differentiated primordial ancestors: "We have thus far been baffled in all our attempts to account for the differences between the races of man; but there remains one important agency, namely Sexual Selection, which appears to have acted as powerfully on man, as on any other animal. I do not intend to assert that sexual selection will account for all the differences between races. An unexplained residuum is left . . . It can be shewn that the differences between the races of man, as in colour, hairyness, form of features etc. are of the nature which it might have been expected would have been acted on by sexual selection" (DM 1:249–250).

Sexual selection gathers together, through attraction and preference, individual variation that, if preferred over many generations, creates categorical variations, which become known as racial variations in the more stark and clear-cut forms that racial differences exist in today. What were once small, possibly biologically insignificant but sexually significant characteristics exert a force in the functioning of sexual attraction, and it is this sexual appeal that gives these otherwise insignificant characteristics a key role to play in inheritance and long-term survival. Sexual selection provides a powerful force in the operations of natural selection. It is ironic that a system as impersonal and as algorithmic as natural selection, at least if Dennett's reading is correct, must deviate itself through sexual selection, in which minute variations of individual taste may have significant effects on subsequent generations. This attractiveness is more difficult to reduce to algorithmic form than Dennett's proposition implies: there is something incalculable about the attraction to others, even in spite of attempted codifications of attractiveness in terms of evolutionary fitness (e.g., the claim, common in sociobiology, that women seek the attractions of older, richer, more successful men, who in turn seek the attractions of younger, voluptuous, reproductively able women, an interesting argument but one that explains only some but not the profusion of sexual relations that make up a culture). This attractiveness links the phenotypic body, more than the active, successful genotype, to an irreducible role in sexual reproduction: "It is certainly not true that there is in the mind of man any universal standard of beauty with respect to the human body. It is, however, possible that certain tastes may in the course of time become inherited, though I know of no evidence in favour of this belief; and if so, each race would possess its own innate ideal standard of beauty" (DM 2:353–354).

Darwin is rarely very specific about the linkage between sexual selection and the development of racial differences, but his strongest hint is this claim about culturally specific standards of beauty: different groups of men and women find different characteristics beautiful or attractive, and these tastes may be inherited. If this is the case, then sexual selection may work on the manifest or visible elements that distinguish one race from another, such as skin color and other bodily characteristics. Although there is no direct proof of this, Darwin hypothesizes that it is these variable standards of beauty and taste that are responsible for the increasing divergences of races from each other: "The best kind of evidence that the colour of the skin has been modified through sexual selection is wanting in the case of mankind; for the sexes do not differ in this respect, or only slightly or doubtfully. On the other hand, we know from many facts already that the colour of the skin is regarded by the men of all races as a highly important element in their beauty; so that it is a characteristic which would be likely to be modified through selection . . . It seems at first sight a monstrous supposition that the jet blackness of the negro has been gained through sexual selection; but this view is supported by various analogies, and we know that negroes admire their own blackness" (DM 2:381–382).

He claims that the differences between different human races is a later evolutionary occurrence than the emergence of recognizably human progenitors; in other words, racial differences are modifications or variations of a newly emergent human form, not steps in the progressive development from the primate to the civilized European, as much of nineteenth-century British colonial culture assumed: "As the newly-born infants of the most distinct races do not differ nearly as much in colour as do the adults, although their bodies are completely destitute of hair, we have some slight indication that the tints of the different races were acquired subsequent to the removal of the hair, which . . . must have occurred at a very early period" (DM 2:382).

Because racial characteristics emerge more clearly as the infant develops than at birth, and because varying characteristics like skin color are evolutionarily later developments than, say, the loss of body hair, Darwin infers that racial relations postdate and depend on sexual bifurcation, that racial variations are contingent on sexual selection, whose results are submitted to the processes of natural selection. This means that there are two contrary tendencies at work in the genealogy of racial differences: a tendency toward greater differentiation over time, that is, the tendency to recursively magnified selective choices over generations (a hypothesis based on Darwin's claim that it is possible that taste and aesthetics may be heritable characteristics: it is possible that the next generation's males and females will exhibit the same broad tastes as the previous generation); and conversely, the mutability of the boundaries dividing the races sexually, their inherent openness to the *next* reproductive encounter, to sexual encounters with those outside their own group or race. There is a strong probability, given the ever-increasing migration and movement of members of different races, that racially defined or designated characteristics will cross more and more frequently. In short, the historical variability of racial differences is directly tied to the immutability of sexual differences, the ever greater divergence of sexually specific features over time.

Political Implications

To sum up, then, some of the key implications of Darwin's model of inheritance, variation, and descent, of the place of the human in the schema of evolution, and the role of sexual division in it:

1. Darwin's work, if it is relevant at all to the humanities and social sciences, is not only a theory of biological development but also a theory of cultural, social, and political development. What Darwin's work makes clear is that what has occurred to an individual in the operations of a milieu or environment (it matters little here if it is natural or cultural) is the force or impetus that propels that individual to processes of self-transformation through his or her sexual relations and his or her relations of inventive survival in a world of tension and competition, and that propels groups, variants, and species to their self-overcoming through their accommodation to natural selection. The struggle for existence is precisely that which induces the production of ever more viable and successful strategies, strategies whose success can be measured only by the degree to which they induce transformation in the criteria by which natural selection functions. Evolution and growth, in nature as in culture, are precisely about overcoming what has happened to the individual through the history, memory, and innovation open to that individual and his or her group. This is true of the survival of species as much as it is of the survival of languages and of political strategies and positions, historical events, and memories. Darwin makes clear that self-overcoming is incessantly, if slowly, at work in the life of species. Politics is an attempt to mobilize these possibilities of self-overcoming in individuals and groups. The logic by which this self-overcoming occurs is the same for natural as for social forces.

2. Darwin accounts for the social, not directly, not through its reduction to biology, for the biological structure of man in no way preempts the forms of social organization within which he will live, but through the logic of natural selection. Culture cannot be viewed as the completion of nature, its culmination or end, but must be seen as the ramifying product and effect of a nature that is ever prodigious in its techniques of production and selection, and whose scope is capable of infinite and unexpected expansion in directions that cannot be predicted in advance. Language, culture, intelligence,

reason, imagination, memory—terms commonly claimed as defining characteristics of the human and the cultural—are all equally effects of the same rigorous criteria of natural selection: unless they provide some kind of advantage to survival, some kind of strategic value to those with access to them, there is no reason why they should be uniquely human attributes or unquestionably valuable attributes. Darwin affirms a fundamental continuity between the natural and the social, and the complicity, not just of the natural with the requirements of the social, but also of the social with the selective procedures governing the order and organization of the natural.

3. Darwin provided a model of time and development that refuses any pregiven aim, goal, or destination for natural selection. This already serves to differentiate him from virtually all of his followers. He refuses anything like the telos or directionality of the dialectic, or a commitment to progressivism in which we must always regard what presently exists as superior to or more developed than its predecessors. We cannot assume that the goal of natural selection is the survival of the individual or the species, nor can we assume that the goal of evolution is the proliferation of progeny. Darwin makes it clear that many species support and indeed require nonreproductive members; it is thus not clear that any pregiven aim or goal can function as the purpose or goal of evolution.

4. This logic of self-overcoming, which is the motor of Darwinian evolution, must be recognized not only as a distribution of (geographical and geological) spacing, but above all as a form of temporization, in which the pull of the future exerts a primary force. Beings are impelled forward to a future that is unknowable, relatively uncontained by the past. It is only retrospection that can determine what direction the paths of development, of evolution or transformation, have taken, and it is only an indefinitely deferred future that can indicate whether the past or the present provides a negative or positive legacy for those that come. This means that history and its cognate practices (geology, archaeology, anthropology, psychoanalysis, medical diagnosis, etc.) are required for understanding the current, always partial, and residual situation as an emergence from a train of temporal events already given, which set the terms for, but in no way control or direct, a future fanning out or proliferation that follows directions latent or virtual but not necessarily actualized in the present. But history, the partial and ambiguous record of the past, is not adequate to indicate particular trends, directions, and variations in the future. It tells us only of what has been, not of what can or will be.

5. One of the more significant questions facing contemporary feminist

theory, and indeed all political discourses, is precisely what generates change, how change is facilitated, what ingredients, processes, and forces are at work in generating the conditions for change, and how change functions in relation to the past and the present. Darwin presents, in quite developed if not entirely explicit form, the germs for an account of the place of futurity, the direction forward as the opening up, diversification, or bifurcation of the latencies of the present, which provide a kind of ballast for the induction of a future different but not detached from the past and present. The future emerges from the interplay of a repetition of cultural/biological factors and the emergence of new conditions of existence: it must be connected, genealogically related, to what currently exists, but is capable of a wide range of possible variation or development of current existence. The new is the generation of a kind of productive monstrosity.

6. Darwin provides feminist and cultural theory with a way of reconceptualizing the relations between the natural and the social, between the biological and the cultural, outside the dichotomous structure in which these terms are currently enmeshed. Culture cannot be seen as the overcoming of nature, as its ground or necessary mode of mediation. According to Darwinian precepts, culture is not different in kind from nature. Culture is not the completion of an inherently incomplete nature (this is to attribute to man, to the human and to culture, the position of evolution's destination, its telos or fruition, when what Darwin makes clear is that evolution is not directed toward any particular goal and has no destination).[15] Evolution produces variation for no reason; it values change for no particular outcome; it experiments, but with no particular results in mind; it has prolific means but no ends. It is an excessive productivity, with each culture an expression of its excessive and multiple possibilities of transformation and elaboration, each culture a surprise to and a development of nature itself.

7. Darwin's work may add some welcome layers of complexity to understanding the entwinement of relations of sexual and racial difference, which can no longer be considered either independent but crossing terms, or additive terms. His work makes clear how sexual selection, that is, relations of sexual difference, may have played a formative role in the establishment of racial differences and, moreover, how racial variations have fed into and acted to transform how sexual difference, subjected to the laws of heredity, is manifested. Darwin provides an ironic and indirect confirmation of the Irigarayan postulation of the irreducibility of sexual difference and its capacity to play itself out in all races and across all modes of racial difference.[16] His work indirectly demonstrates the way that racial and other bodily differ-

ences are bound up with and complicated by sexual difference and the various, transforming criteria of sexual selection.

8. Darwin's work, with the centrality it attributes to random variation, to chance transformations, and to the unpredictable, has provided something of a bridge between the emphasis on determinism that is so powerful in classical science and the place of indetermination that has been so central to the contemporary form of the humanities. Evolution is neither free and unconstrained nor determined and predictable in advance. It is neither commensurate with the temporality of physics and the mathematical sciences, nor is it unlimited in potential and completely open in direction. Rather, it implies a notion of overdetermination, indetermination, and a systemic openness that precludes precise determination.

9. Darwin provided a model of history that is fundamentally open but also regulated within quite strict parameters. There are historical constraints on what becomes a possible path of biological/cultural effectivity: it is only that which has happened, those beings in existence, now or once, that provide the germs or virtualities whose divergence produces the present and future. That which has happened, the paths of existence actualized, preempt the virtualities that other existences may have brought with them; they set different paths and trajectories than those that might have been. History is a broader phase space than that which can be occupied by living beings. And the history or genealogy of living beings transforms and magnifies this phase space, the space of virtualities or latencies, as they transform themselves. While history remains open-ended, the past provides a propulsion in directions, unpredictable in advance, which, in retrospect, have emerged from the unactualized possibilities that it yields.

PART II. NIETZSCHE AND OVERCOMING

4. NIETZSCHE'S DARWIN

Darwin's understanding of natural selection posits both an internal dynamism within living beings and an external force, outside individuals, which generate two parallel series: the series of individual variations and the series of natural selection. Darwin needs both an internal force that produces variations, that induces internal transformations, but gives them no specific direction or purpose, and a rigorous external dynamism, an environment composed of other living beings, in coexistence and competition with each other, which undergoes periodic cycles of flood, drought, and other natural cataclysms, material events that are indifferent, blind, to what they function to weed out or eliminate. While these series run in parallel, they invariably but randomly cross, submitting the first series, random variation, to the second series, natural selection, to produce a system for the generation of ever more elaborate variation and difference. The chance changes in individual variation cross with and are evaluated by chance changes, events, in the environment. The less fit are eliminated, leaving open the possibilities for the proliferation of the more fit, the more varied, the "successful" experimental strategies of existence.

I have argued that if the Darwinian model is appropriate for understanding the complexity and variability of biological existence, then it must also provide an appropriate model for cultural life—not by way of analogy, as social Darwinism proclaims when it argues that memes are "like" genes or ideas are "like" biological entities, but directly, through the same principles and processes, through the dynamic movement of elaboration that coheres in temporal emergence. In culture, too (and partly as a result of the growing complexity of biological existence), there is the infinite but commonly insignificant production of variation and the operation of various selective criteria for evaluating what counts as provisional fitness or successful adaptation. Darwin produced not only an account of the development and continuous refinement of biological existence that is provisionally adapted to its environment by the active forces exerted by that environment; he inadvertently provided us with a theory of time which regards it as an always open-ended movement directed to a future whose parameters cannot be known in advance but whose conditions always and only exist through the continuity between the past and the present. It is the active work of time itself, time as rewriting, time as overlay, development, cohesive succession, that defines the underlying logic of these two series, the short-term time span of reproduction directing individual variation and the long-term time scale of the climatological and environmental changes most central to natural selection, so that they transform each other, so that they dynamize and come into direct contact with each other, so that natural selection modulates individual variation, while the more successful forms of individual variation also transform the landscape of natural selection.[1]

Each "interferes" with the other, each series resonates and transforms the other. History, both natural and cultural history, is the accumulation of a past that preempts the paths but doesn't predetermine them, that generates the future. The future is built on what of the past has survived into the present and with the virtual or potential left in those that do not survive; there is no other future, for Darwin, than that prefigured and made possible by but not contained in the terms of the present. The future brings what has been led up to but has never been defined by the past. Darwin makes the concept of emergence central not only to his conception of life, but to all of its effects, its products.

In this chapter, I turn to what indirectly, surprisingly, "evolved" from Darwin's own model of natural selection: not Darwin's own progeny, the scientists—biologists, embryologists, cognitive theorists, ethologists, and psychologists—who followed his meticulous empirical pathway of observational research, but his more unfaithful heirs, his wayward children: the philosophers troubled by his writings, the unexpected transformations, evolutions, or becomings, that his work undergoes in the hands of Friedrich Nietzsche and, in subsequent chapters, Henri Bergson, what we might understand as his illegitimate philosophical heirs. Nietzsche produces a kind of monstrous misreading of Darwin, a reading largely based on many secondary versions of Darwin, that is, a reading mediated by others' readings of

Darwin's dangerous idea, which nevertheless comes to serve as a mutating mechanism, a teratological agent, in expanding out Darwin's work to include more than is readily palatable for or tolerable by prevailing scientific interests. Nietzsche introduces a properly conceptual dimension to Darwin's own meticulous concern with details and facts, counterposing against Darwin's scientific timidity his own bold and sometimes soaring flights of philosophical conceptualization. Nietzsche enables us to consider the most abstract elements of Darwinism, Darwin at his most philosophical and political, even if Nietzsche does not provide a reading of Darwin himself, of his own texts or precise claims.

Darwin's Grayness

In this chapter, I explore Nietzsche's productive (mis)reading of Darwinism, the objections and criticisms that he formulates regarding what he understood as Darwin's work.[2] In the following chapter, I look in some detail at the ways his transformation of Darwin leads to his own reformulation of the movements of temporality in his conceptualization of history and the past as that which must be active and of use to the present and future; in chapter 6, I examine the ways his conceptions of past, present, and future are bound up with an account of the eternal return that may well serve to revitalize and redirect Darwinism, to propel it, in spite of itself, into a philosophy of existence conceived as a fundamentally open but also constrained multiplicity of becomings, of as much relevance to cosmology, culture, and politics as it is to biology. Through its Nietzschean inversions, Darwinism may come to provide a philosophically based explanation of the planetary universe and its history, as well as the structures of political power, with the same broad conception of temporal openness that marks Darwin's understanding of biological emergence.

Three facets of Nietzsche's writing—his critique of Darwinism, his understanding of history, and his conception of the eternity of time, of time's infinity (really only a fragment of a reading of certain propositions Nietzsche signals but leaves unelaborated in his writings)—are linked to each other through the centrality Nietzsche accords to the concept of the will to power, which both transforms Darwin's concept of the struggle for existence and anticipates Bergson's conception of vital force. I devote a chapter to each of these concerns.

In this chapter, I explore not the well-timed, the well-adapted, the genesis and "perfection of design" of man and beast (what Dennett regards as

Darwin's primary objects of investigation), but rather, the untimely, the dislocated, that which precedes, surpasses, and moves beyond man, that which goes beyond the human and unhinges progress and continuity, displacing the known and the present for a future that does not yet exist. As Nietzsche says, "I write for a species that does not yet exist" (1968b, *The Will to Power* [hereafter WP], #958). Through Nietzsche, we can detect a certain derangement that is at the center of the ordered brutality of the struggle for existence, the emergence of excess, of a beyond-survival, a superfluous remainder that marks life in ways Darwin may have recognized but could not fully appreciate. To Darwin's conception of life as survival Nietzsche adds the joyous excess or superfluousness of inner force, the political and moral force, the rare magnificence, of the will to power.

Although *The Origin of Species* was published in 1859, when Nietzsche was fifteen, the writings of Nietzsche and Darwin's later works are close in publication dates. Quite fervent discussions around Darwin's writings must have occurred throughout the period of Nietzsche's studies, and Darwinism remained one of the most contested and elaborated theoretical contributions of the nineteenth century. *The Descent of Man* was published in 1871, and *The Birth of Tragedy* was published in 1872. It seems quite clear from what Nietzsche does say about Darwin that he did not read Darwin's own texts, but instead was familiar with some of Darwin's sources (Malthus and Lamarck), with some of the writings of the English Darwinians (Herbert Spencer and T. E. Huxley), and with quite a number of German Darwinians (among them, Ernst Haeckel), with whom he quite justifiably disagreed. His reading seems to uncannily and critically anticipate and sidestep the development and revitalization of Darwinism under the rubric of social Darwinism, where it becomes a reductive account that explains the development of social, sexual, and political life in terms of strategies of reproductive success. For this reason alone, as an antidote to the spirit of scientism that marks so much of contemporary Darwinism, it is worth exploring further what Nietzsche's concerns about Darwinism might be and how integral these are to the development of his own affirmative philosophy of force.

Arthur Danto (1968, 224), one of Nietzsche's more distinguished commentators, has remarked, "It is difficult to justify the title of so many of Nietzsche's aphorisms which are headed 'Anti-Darwin,' or words to that effect. As with much of the discussion of Darwinism, his polemic was ideological rather than scientific, and it had scant bearing on the true interest and importance of Darwin's theories. Strictly speaking, then, there is no excuse for an extended discussion of Nietzsche's views of Darwin. He had some private image he thought to be Darwin."

More contemporary readers of Nietzsche, such as Keith Ansell Pearson (1997b, 90), in his book *Viroid Life*, confirm Danto's claim: "Nietzsche is responding to Darwinism, not so much as a biological theory but more as a social theory, as *social* Darwinism."[3] Keeping in mind this insight that Nietzsche had not read Darwin's own writings, some of his objections may not be as appropriate to Darwin's own texts as they are to his followers, interpreters, and detractors. In other words, though Nietzsche has an intriguing and complex relation to evolutionary theory, we should not confuse his arguments and claims with a critique of Darwin himself. Nietzsche's reading is an important one, a necessary corrective to the impulses behind the clumsy importation of Darwin into social, cultural, and political thought, as long as we keep in mind to whom his criticisms should be properly directed. Though Nietzsche's analysis is to some extent misdirected, nevertheless it can itself be interpreted as a mutation, a particularly sturdy and resilient, if rare, variant of Darwinism that may direct some Darwinian interests in new, unexpected directions, away from its monopoly by the empirical sciences and into unexpected conjunctions with ontology and morality.

Much of Nietzsche's conception of Darwinism is compressed into a single proclamation in *The Gay Science*, which can be filled out, or at least partially explained by connecting it to a series of other, briefly developed claims Nietzsche makes, scattered here and there throughout his published and unpublished writings. Let us begin with his proclamation, which will help demonstrate some of the flavor of his intervention into Darwin's tamer world:

> There are truths that are recognized best by mediocre minds because they are most congenial to them; there are truths that have charm and seductive power only for mediocre spirits: we come up against this perhaps disagreeable proposition just now, since the spirit of respectable but mediocre Englishmen—I name Darwin, John Stuart Mill and Herbert Spencer—is beginning to predominate in the middle regions of European taste. Indeed, who would doubt that it is useful that *such* spirits of a high type that soar on their own paths would be particularly skillful at determining and collecting many common facts and then drawing conclusions from them: on the contrary, being exceptions, they are from the start at a disadvantage when it comes to the "rule." Finally, they have more to do

> than merely gain knowledge—namely, to *be* something new, to *signify* something new, to *represent* new values. Perhaps the chasm between *know* and *can* is greater, also uncannier, than people suppose; those who can do things in the grand style, the creative, may possibly have to be lacking in knowledge—while on the other hand, for scientific discoveries of the type of Darwin's a certain narrowness, aridity and industrious diligence, something English in short, may not be a bad disposition. (1974 [hereafter GS], 5 #349)

Here, in his typically acerbic and elliptical manner, Nietzsche develops a number of claims and assumptions that are also articulated elsewhere in his work. I use this quotation to provide an outline of some of the strands and implications knotted together here in the form of a broad and incomplete list, largely because these claims are not systematically developed in his other works. These are reconstructions of hints and suggestions that he makes largely in passing:

1. There is a strong link for Nietzsche between liberalism, economism, and Darwinism. These discourses (to be more specific, those links between, say, Darwin, Malthus, John Stuart Mill, Adam Smith, David Ricardo, and others) must be understood as servile in two senses. First, each involves a distinctively "English" mode of pettiness, a concern with equalization, which, for Nietzsche, is always a mode of downward descent, reducing man to a common denominator, the lowest level, the herd. Egalitarianism reduces all to the mediocre, the average. He implies that this peculiarly English malaise is perhaps an effect of the reduction of the English working class to a form of indistinct replaceability, and of the middle class to apologists for this process of downward leveling, of which they themselves are an effect. Second, these uniquely English discourses function by the accumulation of tiny details, by a servility to facts, political or empirical, that limits any possibilities of conceptual or philosophical innovation and creation. Darwinism and liberalism both function with their noses to the ground, looking downward, and never with their heads in the sky, soaring above, high with aspirations. Darwinism and liberalism are discourses of shame, self-abnegation, pettiness. They are mired in administration, in ordering, compiling, filing; they are fundamentally bureaucratic. They have no grandeur of vision, no aspiration to overcome what is here and now. They remain "arid," "industrious," careful, and uninventive.

2. Darwinism, Nietzsche argues, is thus a discourse of the triumph of the weak over the strong, the herd over the individual, the servile over the noble,

the low over the high, the mediocre and the average over the exceptional and the strong. The struggle for existence ensures that, for Nietzsche, paradoxically, it is the unfit who survive and the fit who become extinct. It is the struggle of the unfit rather than the fittest. The movement of evolution is the movement away from instincts, the primitive and noble directness of the lion and other beasts of prey toward that encumbrance that is consciousness, a movement of self-cultivation that induces conscience and reflection. These are movements away from the form of the warrior, the risk taker, the creator: evolution produces bourgeois man, man as a tiny cog in a giant machine, as its inevitable, if passing, consequence. What Nietzsche admires is not so much a survival of the fittest, the norm, the well-adapted, the boring, as a survival of the noblest, of the exceptional, even the monstrous, the one who can bear to overcome, to become unrecognizably more. He elevates as the highest form of evolution the production of the untimely individual, one not so much adapted as an exaptation, nonadapted, out-of-place, the one secreted, reveling, in life's luxurious surplus: "*Spiritual enlightenment* is an infallible means for making men unsure, weaker in will, so they are more in need of company and support—in short, for developing the *herd animal* in man . . . The self-deception of the mass concerning this point, e.g., in every democracy, is extremely valuable: making men smaller and more governable is desired as 'progress'!" (WP 1 #130).

Evolution privileges the weak, the average, the mediocre, and the herd over the exceptional individual, and this accounts for its gray and somber character. This may be why, contrary to Danto, it makes sense that Nietzsche commonly represents himself, not only as the Anti-Christ, but also as the Anti-Darwin, the champion of the exceptional, the unique, the unrepeatable, the advocate for what overcomes the expected and hoped for to make and be something new:

> What surprises me most when I survey the broad destinies of man is that I always see before me the opposite of that which Darwin and his school see or want to see today: selection in favor of the stronger, better-constituted and the progress of the species. Precisely the opposite is palpable: the elimination of the lucky strokes, the uselessness of the more highly developed types, the inevitable domination of the average, even the *sub-average* types. If we are not shown why man should be an exception among creatures, I incline to the prejudice that the school of Darwin has been deluded everywhere.
>
> That will to power in which I recognize the ultimate ground and char-

acter of all change provides us with the reason why selection is not in favor of the exceptions and lucky strokes: the strongest and most fortunate are weak when opposed by organized herd instincts, by the timidity of the weak, by the vast majority. My general view of the world of values shows that it is not the lucky strokes, the select types, that have the upper hand in the supreme values that are today placed over mankind; rather it is the decadent types—perhaps there is nothing in the world more interesting than this *unwelcome* spectacle. (WP 3 #685)

The Darwinian picture that Nietzsche addresses is one of unremitting mediocrity, averageness, the privilege of the tiny, the unexceptional, the herd or group. If there is an evolutionary push forward, a movement of becoming that is also a self-overcoming, for Nietzsche this is the prerogative only of the most brave and fearless, the most noble of individuals, only of the human who joyously seeks beyond the human, who seeks a kind of happy self-annihilation as human, the one whose laughter makes Zarathustra weep joyously.

The Noble and the Servile

Nietzsche claims that Darwinism elevates the struggle for mere existence, for need or survival above the struggle for something more noble, which is the struggle that exhibits and strengthens the will to power.

3. The Darwinian model is based on lack, on scarcity, on the absence of adequate provision for all, the very model that has governed all representations of desire itself. In place of this Malthusian absence or shortage that models and motivates all Darwinian activity (the struggle for existence is conceived largely as the struggle to avoid death, but is never conceived as a struggle to enjoy life), Nietzsche wants to suggest a superabundance or excess of energy, of force, of resources, and of production. It is this extra, this surplus over and above need or requirement, that enables or induces self-overcoming, that creates the exceptional, and that is the goal of Dionysian celebration: "As regards the famous struggle for existence, it seems to me that this is asserted rather than proved. It takes place, but it is the exception. The general aspect of life is *not* need, nor starvation, but far more richness, profusion, even an absurd prodigality. Where there is struggle, there is struggle for power" (1968a, *Twilight of the Idols* [hereafter TI], 9:14).

The struggle for existence is simply the servile struggle for life, but it cannot engender that luxury or abundance that is necessary for the production of all that is higher and all that is necessary for that movement of self-overcoming that Nietzsche values so highly. The struggle for existence aims too low: it aims only for existence, for bare survival, for mere life itself. For Nietzsche, though, what ennobles, what elevates, is the struggle for power, the struggle to overcome what one can do and what one is, the struggle to command and, in commanding, to invent. The struggle for existence is what we share, but the will to power that we actively embrace differentiates us and elevates us above ourselves. Moreover, Nietzsche makes it clear that even the most elementary living creatures are not content with mere existence: the struggle for more—that is, the will to power—is greater than the struggle for existence, for it aspires to more and it thus accomplishes more. In any case, this struggle for self-preservation, for survival, is by definition a losing fight. If survival is the goal of life, life fails in every case! "One cannot ascribe the most basic and primeval activities of protoplasm to a will to self-preservation for it takes into itself absurdly more than would be required to preserve it; and above all, it does not thereby 'preserve itself,' it falls apart. The drive that rules here has to explain precisely this absence of desire for self-preservation" (*WP* #651).

4. Throughout his work, Nietzsche attempts to substitute the will to power for the power of individual variation. In the process, he aims to render the biological being more active and the environment more reactive, a reversal of his view of the Darwinian reduction of the living being to a reaction to the external selective pressures imposed by the environment. Life, existence, is about, and the play of, the will to power, the will to command or to obey, which is a will to proliferation and profusion rather than an imperative to simply live or to merely reproduce. This is precisely his position in *The Gay Science*:

> That our modern natural sciences have become so thoroughly entangled in the Spinozistic dogma [of the centrality of self-preservation] (most recently and worst of all, Darwinism with its incomprehensibly one-sided doctrine of the "struggle for existence") is probably due to the origins of most natural scientists: In this respect they belong to the "common people"; their ancestors were poor and undistinguished people who knew the difficulties of survival only too well at firsthand. The whole of English Darwinism breathes something like the musty air of English overpopulation, like the smell of distress and overcrowding of small people. But a natural scientist should come out of his human nook; and in nature it is not conditions of distress that are *dominant* but overflow and squandering,

> even to the point of absurdity. The struggle for existence is only an *exception*, a temporary restriction of the will to life. The great and the small struggle always revolves around superiority, around growth and expansion, around power—in accordance with the will to power which is the will of life. (GS #349)

It is not "distress," "scarcity," "survival" that characterize nature, but abundance, overflow, and profusion. Rather than advocating an economy of counting, observing, administering, Nietzsche proclaims an economy of excess, a general rather than a restricted economy, in Bataille's sense (1985). Life is not about mere survival, but about profusion and proliferation, not existence but excess, not being but being-more, that is, becoming, but a becoming-what that cannot be determined in advance, that is always itself in the process of becoming-something-else.

5. Darwin's theory of evolution, much like Mill's understanding of utilitarianism (the greatest happiness of the greatest number, the averaging of happiness as the plan for a universal ethics), places adaptation and utility at the forefront. But adaptation, for Nietzsche, is always the melding of an internal will to an external force, the making passive of life relative to the nonliving forces outside of life. It is the taking for granted of an environment, and the adaptation of life to that environment, rather than the utilization and transformation of environments by a life now configured as active:

> There is no more important proposition for all kinds of historical research than that which we arrive at only with great effort . . . namely, that the origin of the emergence of a thing and its ultimate usefulness, its practical application and incorporation into a system of ends, are *toto coelo* separate; that anything in existence, having somehow come about, is continually interpreted anew, requisitioned anew, transformed and redirected to a new purpose by a power superior to it . . . everything that occurs in the organic world consists of *overpowering, dominating*, and in their turn, overpowering and dominating consist of re-interpretation, adjustment, in the process of which their former "meaning" and "purpose" must necessarily be obscured or completely obliterated. (1967, *On the Genealogy of Morals* [hereafter GM], 2 #12)

Usefulness, utility, is a side effect, an extra bonus, of the articulation and expression of the will to power. The useful is always subservient to the requirements of the present: it lacks any element of the untimely. The useful, the utilitarian, the adapted is always that which exploits external circumstance, the given, but which cannot make the given anew, cannot make itself something other. The useful performs only in light of the actual, the real, the present, but it has no aspiration to the untapped, the unknown, the new that attracts and elevates life, even in its animal forms: " 'Useful' in respect of acceleration of the tempo of evolution is a different kind of 'useful' from that in respect of the greatest possible stability and durability of that which is evolved" (WP 3 #648).

In other words, the useful can only be understood in terms of the privilege of the present and the known. The useful is what is at hand, available for use now. Nietzsche himself, though, is concerned with a different—deranged—sense of usefulness, the useful in nondeterminable contexts, the useful for a future that cannot be predicted, in other words, a useful that is not so much of use but awaiting the invention of a use from its current excess. It is the result of an excess in present usefulness. The true "usefulness" of Darwin's account is domination, force, that which enables a being to overcome:

> "Useful" in the sense of Darwinist biology means: proved advantageous in the struggle with others. But it seems to me that the feeling of increase, the feeling of becoming stronger, is itself, quite apart from any usefulness in the struggle, the real *progressus*; only from this feeling does there arise the will to struggle—.
>
> Physiologists should think again before positing the "instinct of preservation" as the cardinal drive in an organic creature. A living thing wants above all to *discharge* its force: "preservation" is only a consequence of this.—Beware of *superfluous* teleological principles! The entire concept "instinct of preservation" is one of them. (WP 3 #649–650)

The instinct of preservation, the production of the useful, what facilitates survival, is always about an acceptance of the circumstances, the situation in which one finds oneself. It involves a quietism, a passivity relative to one's environment, the very passivity, the acceptance of the given, Nietzsche believes, of the Darwinian subject relative to its surroundings. Nietzsche seeks instead a will to overcome, an active and internal will to make more, a will not to find, to see, to discover, or to accept, but a will to produce, to make, more and other. The useful is only what can be considered useful here and now, in an understood or inferred present: what Nietzsche instead seeks is a useful in a context yet to be determined, an untimely kind of utility, useful for those who do not yet exist or are yet to come, useful in unimaginable contexts.

6. Nietzsche also implies, in a manner that is suspicious of, if not hostile

to, the scientific spirit engendered by Darwin's own researches, that there is now and always has been a rift between doing/being on the one hand, and knowing on the other, an opposition between ontology and epistemology that has functioned in the service of science. For him, there are two contrary forces, one expanding the sphere of possible actions and thus enhancing the will to power (doing), the other contracting and limiting these actions, muting them, regularizing them, making them understandable and thus limiting their force while being more able to control their effects (knowing). This rift between knowing and doing, between epistemology and ontology, is precisely what Darwin sought to avoid when he affirmed the positivity of the real in the generation of life and of its evolutionary outcome, intelligence, but that Bergson, as we will see, suggests is itself an evolutionary, that is, historical consequence of the emergence of intelligence, the impulse to freeze, to spatialize, to control. It is the scientific impulse to render pragmatic action repeatable, and to increase its differentiation from the movements of intuition, the impulse to be within, to act and make among and as things: "Physiologically . . . science rests on the same foundation as the ascetic ideal: a *certain impoverishment of life* is a presupposition of both of them—the affects grown cool, the tempo of life slowed down, dialectics in place of instinct, seriousness imprinted on faces and gestures (seriousness, the most unmistakable sign of a labored metabolism, of a struggling laborious life)" (GM 3 #25).

7. Perhaps most significant, Nietzsche wants to claim that Darwinism misunderstands the will to power which animates all of life from *within*. The history of life, as readily as the history of inert matter itself, is the history of the activity of the will to power, the internal force of expansion that makes living beings become and act, rather than exist as and through an essence or a past. This active internality, which is why Nietzsche uses the term *will* to describe it, is the motor of history, and also of the environment, the existence of other species and individuals acting as modes of provocation for greater self-overcoming and expansion of these internal forces, which strive against each other as well as against external forces: "The influence of 'external circumstances' is overestimated by Darwin to a ridiculous extent: the essential thing in the life process is precisely the tremendous shaping, form-creating force working from within which *utilizes* and *exploits* 'external circumstances'—The new forms molded from within are not formed with an end in view; but in the struggle of the parts a new form is not left long without being related to a partial usefulness and then, according to its use, develops itself more and more completely" (WP 3 #647).

In short, Nietzsche regarded Darwin, and Darwinism, as a reactive discourse, which privileges the weak, the average, the external, and ties the movement of history to the least discernible and most readily assimilable change, that links progress to the ever more useful development of organs and aptitudes and that seeks emersion in, acceptance of, and adaptation to a given environment. In place of this reactive view, Nietzsche seeks an active notion of life, one in which not adaptation but expansion has a privileged role, and where the future designates the time, not of man, but of the overman. The overman cannot be the product of natural selection, but is the consequence of artificial selection, though an artificial selection engineered by forces higher than human breeders—the breeding of the superior by means of the eternal return.

In place of natural selection, Nietzsche positions the eternal return, a view of history as the triumph of the strong, not, as Nietzsche sees Darwin, a history of the triumph of the herd, the species, nor, as Dawkins suggests, the statistical majority: "My philosophy brings the triumphant idea by which all other modes of thought will ultimately perish. It is the great cultivating idea: the races that cannot bear it stand condemned; those who find it the greatest benefit are selected for mastery" (*WP* #1053).

Nietzsche wants to bring forth, from Darwin's own heritage, an activity that Darwin did not recognize adequately: the active force of a seizing hold of a thing, of matter. Life is not a reaction to matter or nature, but a seizing of matter and a rendering it for one's own purposes to come. Life is not simply a subjective interpretation imposed on matter that leaves matter absolutely intact and untouched; interpretation is itself the activity of making matter over, overcoming its form by one's own forces, the wills to power that compete to make life more than itself. Nietzsche understands clearly that evolution is overcoming, and that evolution is thus a process that, if it occurs at all, occurs equally for ideas, customs, and social practices as much as it does for organs, morphologies, and instincts: "The 'evolution' of a thing, a custom, an organ is thus by no means its *progressus* toward a goal, even less a logical *progressus* by the shortest route and with the smallest expenditure of force—but a succession of more or less profound, more or less mutually independent processes of subduing, plus the resistances they encounter, the attempts at transformation for the purposes of defense and reaction, and the results of successful counteractions. The form is fluid, but the 'meaning' is even more so" (*GM* 2 #12).

The form and meaning of matter is what is actively produced through life's encounter with matter, including the (biological) matter of life itself.

Evolution is not a movement directed simply from within a given being, an internal unfolding, nor is it the impact of external forces, a making-passive of life; it is a rewriting, a remaking, a seizing of a thing, an obstacle, a given, by something new, the force of life, the will to power, and its transformation outside of itself. Evolution, for Nietzsche, designates the precedence of a future that always overwrites and transforms the present, that directs the present to what is beyond its containment. It is not, as Darwin conceived, the reconstruction of the past that helps explain our present, but an understanding of our present, and its dislocations, that helps bring about unknowable futures:

> The cause of the origin of a thing and its eventual utility, its actual employment and place in a system of purposes, lie worlds apart; whatever exists, having somehow come into being, is again and again reinterpreted to new ends, taken over, transformed, and redirected by some power superior to it; all events in the organic world are a subduing, a *becoming master*, and all subduing and becoming master involves a fresh interpretation, an adaptation through which any previous "meaning" or "purpose" are necessarily obscured or even obliterated. However well one has understood the *utility* of any physiological organ (or of a legal institution, a social custom, a political usage, a form of art or a religious cult), this means nothing regarding its origin. (GM #77)

Life, dynamism, force entail that the origin of a thing is fundamentally different from its genealogy, its history and development. This history is composed of a complex of forces, both interpretations and actions, that have seized control of the thing and made it their own. Just as biological descent is not the unfolding of a plan through many generations of obedient offspring, life struggles violently against materiality, and must take on the guise of materiality in order to master matter even momentarily (GM 3 #25). Conversely, matter itself is made up of teeming forces, gravitational, inertial, mechanical, electrical, and so on, whose alignments make life possible and render it dynamic through its forms and energies, through the resources the living extract from it. For Nietzsche, life is overcoming matter, making matter over in one's image, but also succumbing to matter, bending oneself to it, using and being used by it. Utility is always a compromise notion: it covers over the ambiguity of whose use is in question.

Life cannot be understood, then, as the attainment of growing utility, the perfectibility of adaptation, the increasing ingenuity and efficiency of biological design, as Dennett suggests. Rather, it must be understood as the will to ever greater power, the will or multiplicity of wills, that contest their strength, that overcome themselves and each other to multiply their force without direction or a more specific goal. Life is the movement of the stronger, the more powerful, the more dynamic to overcome itself and become other, more, beyond, and to take with it the reactive, the less dynamic in order to convert their forces into its own:

> The case is the same even within each individual organism: with every real growth in the whole, the "meaning" of the individual organs also changes; in certain circumstances their partial destruction, a reduction of their numbers (for example, through the disappearance of intermediary members) can be a sign of increasing strength and perfection. It is not too much to say that even a partial *diminution of utility*, an atrophying and degeneration, a loss of meaning and purposiveness—in short, death—is among the conditions of an actual *progressus*, which always appears in the shape of a will and a way to *greater power* and is always carried through at the expense of numerous smaller powers. The magnitude of an "advance" can even be measured by the mass of things that has to be sacrificed to it; mankind in the mass sacrificed to the prosperity of a single *stronger* species of man—that *would* be an advance. (GM #78)

For Nietzsche, it is not that Darwin is mistaken and needs correction. Nietzsche does not deny or refuse Darwin's analysis of natural selection; rather, he suggests that Darwin's "interpretation" doesn't go far enough, that it fails to see the force, the activity, the power of overcoming that signals an excess of survival, survival's luxurious profusion. Survival, fitness, adaptation are not forms of accommodation, forms of compliance with one's surroundings and one's fellow beings, but modes of striving, dominating, commanding, ordering and obeying, succumbing and submitting. Nietzsche claims that the Darwinians, perhaps even anticipating as contemporary a figure as Dennett, are content to accept the absolutely chance nature of the universe, but they are not willing to accept that this chance has an inner force, a will, an indeterminable direction or interest. What the Darwinian scientists, indeed biologists in general, cannot affirm in refusing the will to power is the force of activity, the primacy of active force, and that this active force itself must be ascribed a (nonsubjective) will, an intentionality:

> I emphasize this major point of historical method [that origin is of less significance than the will to power] all the more because it is in fundamental opposition to the now [and still] prevalent instinct and taste which

> would rather be reconciled even to the absolute fortuitousness, even the mechanistic senselessness of all events than to the theory that in all events a *will to power* is operating. The democratic idiosyncrasy which opposes everything that dominates and wants to dominate, the modern *misarchism* (to coin an ugly word for an ugly thing) has permeated the realm of the spirit . . . Today it has forced its way, has acquired the right to force its way into the strictest, apparently most objective of sciences; indeed it seems to me to have already taken charge of all physiology and theory of life—to the detriment of life, as goes without saying, since it has robbed it of a fundamental concept, that of *activity.* (GM #78)

Darwinism robs life of its activity, or rather, all science robs forces, whether living or not, of their dynamism, their yearning for more. Life at its fullest, at its most forceful, is a dominating, an overcoming, a transformation of those obstacles that stand in its way. It is thus a conversion of resisting or passive forces into its own resources. But science, especially the study of life in the biological sciences, cannot abide by this notion of force, cannot accept its fundamentally antidemocratic, nonegalitarian resonances, cannot accept the differentiating and privileging facility that force represents, that distinguishes active from reactive, dominant from obedient forces: "In all willing it is absolutely a question of commanding and obeying" (1972, *Beyond Good and Evil* [hereafter BGE], #19). The will to power, the desire of each thing, entity, atom, or energy to expand or intensify itself, to become more, is unpalatable to the grayness of the scientific spirit, which seeks only its mastery of the world contained in constraints, in limited conditions, under certain perspectives, and which hides its own investment in rendering this play of forces knowable or, better, predictable and, ideally, controllable:

> Under the influence of the above-mentioned idiosyncrasy [the scientific robbery of activity from life], one places instead "adaptation" in the foreground, that is to say, an activity of the second rank, a mere reactivity; indeed, life itself has been defined as a more and more efficient inner adaptation to external conditions (Herbert Spencer). Thus the essence of life, its *will to power,* is ignored; one overlooks the essential priority of the spontaneous, aggressive, expansive, form-giving forces that give new interpretations and directions, although "adaptation" follows only after this: the dominant role of the highest functionaries within the organism itself in which the will to life appears active and form-giving is denied. (GM #78–79)

The will to power is never singular but always functions with a multiplicity of other wills which may align themselves or function in competition with each other. This is why it cannot be identified with an individual, a commanding subject, a victorious social group or class—in spite of popular readings of Nietzsche. The will to power is sub- or inhuman, prepersonal and impersonal rather than an attribute of a subject, a group, a people, or a race. The will wills the obedience of other wills. Such a will is active and commands, and those wills that obey, that adapt, are reactive.[4] What the scientific spirit seems unable to tolerate is its own will to power, its own reactive requirement of the freezing of (biological or temporal) becoming into the form of knowable being, determinable identity, predictable structure—the impulse of much contemporary Darwinism.

Nietzsche serves both as a monstrous offspring of Darwinism—a Darwinian in Darwin's own conception (if not in Nietzsche's), substituting the will to power for the organic struggle for survival—but also as a corrective to contemporary Darwinism, which has sought, beyond the boldness of Darwin's own conjectures, a security that the knowledge of the past and present will preempt and provide us with knowledge of the future, that time is regular, predictable, knowable. What Nietzsche makes clear is that such a knowledge is possible only with the freezing, the arresting of the active dynamism of the will to power, that is, with freezing and thus killing life itself. It is, in this sense, the most reactive of forces, the ascetic impulse, that eliminates all that is joyous, redundant, unuseful, or irrational.

It is ironic, though, that much of what Nietzsche proclaims as part of his critique of Darwin and Darwinism is consonant with Darwin's own position. As I have already argued, Darwin does not privilege man as the end or goal of evolution; he refuses any telos, aim, or function to evolutionary processes; and he affirms the unpredictability of evolution's movements. He does not make the instinct of self-preservation the primary drive; on the contrary, self-preservation is a bonus for skills and abilities that have been rendered of value by the operations of natural and sexual selection. Nor does he understand fitness as either adaptation to a given environment or the generation of many progeny. In fact, his position is remarkably close to Nietzsche's, in spite of its "Englishness." There is no goal to evolution, and no one element of its operations can serve to explain it. It is not, of course, that Darwin can be assimilated to Nietzsche, but rather that many of the facets of Darwin's work that worry Nietzsche are perhaps elements where Darwin is closest to Nietzsche's own interests.

My goal here has not been to indicate how Darwinism may assimilate the waywardness of Nietzsche's contributions to the study of time and life, nor how Nietzsche may be simply affirmed in his condemnations of Darwin. Rather, I have tried to show that, in spite of some quite remarkable resonances and similarities—their affirmation of the richness of life over the paucity of resources that we have in our knowledge of life, of the fundamental unpredictability of living forms, and their celebration of the centrality of chance, accident, and surprise—they cannot be too closely linked. My goal here has been more to see to what extent Nietzsche can function as an irritant for Darwinism, to what extent he brings out what in the Darwinian tradition is not readily observed: its own moments of hesitation, its reversions to passivity, its refusal to openly acknowledge and provide values by which life lifts itself beyond itself. In the next chapter, I look in more detail at Nietzsche's more positive contribution to a refiguring of time and life as active goals of the will to power, his movement beyond Darwinism, his anticipation of facets of Bergsonism.

5. HISTORY AND THE UNTIMELY

Nietzsche wants to provide a dynamic, active Darwinism, in which life excels, expands, and transforms itself, undergoes or, rather, undertakes, becoming-other and becoming-more. Life is not to be conceived in isolation, alone and independent, without external impetus; rather, life is that which makes what is external part of its own internal dynamic. Nietzsche wants to produce an evaluative, that is, a moral Darwinism, a knowledge of life that privileges the *values* of life itself. But the life that Nietzsche privileges is not mere life, bare and simple existence, but life experienced in its fullest as active force, life that resists prediction and precise calculation, life that remakes itself and its world, life that is victorious. He is concerned with life at its highest, its most active and intense. This concern with the value of the higher life, of a life beyond mere life, develops in Nietzsche's writings through his preoccupation with the question of time and emergence.

The most central element in Nietzsche's conception of life is the question of time, time as that which connects, runs through, things and processes and conjugates itself as past, present, and future. He is interested in how conceptions of the past affect or relate to our concerns in the present and future, with the question of history, with which he begins his erstwhile explorations of time. Time, however, cannot be reduced to history. History is that mode of writing of the past that makes it accessible to and relevant for the present (we must not confuse the two: the past and the study of the past that is history, which is always a past in light of the present). Nietzsche wants to develop a history, a reading of the past, not just in light of the present, but for the future, a history that jumps the intermediary of the present, which constrains it to the known and the useful in order to allow it to function in a way that enervates and welcomes the future. He wants to develop a concept of history that engenders action, not reflection, a history that produces the active rather than justifies the reactive.

His conception of history, and of the past, is an attempt to problematize the prevailing views on which the study of the past had been previously based. This pervasive conceptualization of the past, which still lies today behind positivist historiography, makes several assumptions, among them:

1. The belief that the past is an objective reality that can be more or less accessed and reconstructed in the present through material artifacts and texts retained or gleaned from the past.
2. The belief that the past is much like the present and future; thus, we can learn lessons for our present from a better understanding of the past.
3. The belief that the past is the precondition and thus the (causal) ground of the present, that the present must be seen as a mode of continuity of and coherence with the past.

Like Darwin, Nietzsche is not content to examine the past as the most knowable and efficacious mode of temporality, as the object of justified speculation, insofar as it is "over," finalized, static, and thus "objectively," neutrally, or disinterestedly interpretable. His interest lies not in the past itself but that which in the past still retains its dynamic potential. His interest, along with Darwin's, is directed to providing an understanding of the relations between past and future in which the present is no longer the privileged intermediary, and in which the open-ended becoming of the past into an unknowable future is affirmed. He invents a time that is dynamized, active, unpredictable, a time that does not forget the past but does not find itself constrained to its terms, for the future has a past inherent in it as it veers away from its own causal predictabilities. He invents a time that produces, that makes and unmakes, rather than an inertia, a passive wearing away or mere decay.

Moreover, as few have done before him, he disconnects time from matter, understanding them as two different, independent kinds of forces, sometimes connected and sometimes considered in separation. Time goes on, forever. It is the only material infinity, stretching infinitely into the past and endlessly into the future. Matter, and along with it, life, is finite, limited in its mass, its energy, its force, always outstripped and overcome by the (infinite) passage of time. Time is an active force, not reducible to the forces of matter, which nevertheless confronts both life and matter as their internal limit.

Nietzsche's conception of time is far-flung and suggestive rather than developed, and, for convenience of analysis, can be divided into three components, encapsulated in three broad hypotheses. The first, the hypothesis of the inherence of the past, is based on his understanding of the place of history, that is, the place of present reconstructions, involving as they do both remembrance and forgetfulness. The second, the hypothesis of the generative force of the present, is his understanding of the will to power as the project of the expansion of present forces into the future. These two conjectures occupy the bulk of this chapter. The third, his hypothesis of the eternity of time, is compressed into his enigmatic and misunderstood doctrine of the eternal return; this is the object of investigation in the next chapter.

Memory and Forgetfulness

Nietzsche primarily develops his analysis of the moral and evaluative bases of history in his 1873 essay, "On the Utility and Liability of History for Life" (1995 [hereafter "OUL"]) in *Unfashionable Observations.* Here he presents a critique of the overwhelming power that history and historical studies have exerted on contemporary thought. Nietzsche's purpose, as in much of his work, is to present a "transvaluation of all values," an assessment of the value of value itself, and his exploration of history is precisely an attempt to assess the value of history, its use for the purposes of present and future life. He is concerned, here, as in *On the Genealogy of Morals*, with the question of cultivating a knowledge that affirms life rather than a science that freezes it. It is for this reason that he turns to the question of history, one privileged element, a form of study, of the domain of temporality. History makes itself, lives itself only in the value a future bestows on it. We cannot or should not seek in science the question of the genesis and development of life; it is philosophy that enables us to ask the question of value and of the value of its value. Instead of subordinating history to the interests of science, truth, and objectivity, that is, to the painful accumulation of facts in which Darwin was so adept, Nietzsche seeks in history a way of affirming, ennobling, and cultivating the highest values of life, life at its most enervated, life in its future forms. It is never science that affirms the future, for it, too, like history (after all, it is a disavowed mode of history, a history without knowing it), can only describe the past. Nietzsche's project is to refigure history itself, making history a mode of affirmation of the present and future in which the value of the past can be addressed: "We need [history] for life and for action, not for the easy withdrawal from life and from action, let alone for white-washing a selfish life and cowardly, base actions. We only wish to serve history to the extent that it serves life, but there is a way of practicing history in which life atrophies and degenerates" ("OUL" 85).

Nietzsche aims to think a human sensibility in which history is evaluated, not as a good in itself, not as a preeminent value, but as an aid or discipline that fosters or cultivates our present and future possibilities. It must be recognized that human health and well-being, the capacity to rise above oneself, requires the correct regulation, management, or balance between cultivated memory and a privileged forgetfulness. A being largely dominated by memory and the past is one for whom the present and its possibilities of action are curtailed. To be mired in the past is to be unable to think and act the future; conversely, to be unanchored in the past, to have no connections to or resonances with the past, is also to have no way to see or make a future, it is to have no place from which a future can be made that is different from the present. Well-being requires a judicious mixture of the historical and the ahistorical, the timely and the untimely, the past and the future.

His proposal for the cultivation of history for the purposes of life involves a paradox: for the condition of action, of life, of force involves, above all, forgetfulness, a letting go of memory, the immersion in the immediacy of the present and its nascent possibilities for the future. This is the condition of the animal, and indeed, of the human. Action requires the fullness of the present; it requires forgetting, the capacity to be reborn at each moment without the encumbrance of the past. Nietzsche believes that too strong an immersion in the past is an illness, a debilitation, an inhibition to life, for it prevents our active living in the present. It is this fixation on the past that characterizes *ressentiment* (he who suffers from ressentiment cannot digest the past and be rid of it: it returns to haunt the present). History and life exist in tension: the more there is of the one, the less there can be of the other:

> All action requires forgetting, just as the existence of all organic beings requires forgetting, just as the existence of all organic things requires not only light, but darkness as well. A human being who wanted to experience things in a thoroughly historical manner would be like someone forced to go without sleep, or like an animal supposed to exist solely by rumination and ever repeated rumination. In other words, it is possible to live almost without memory, indeed, to live happily, as the animals show us; but without forgetting, it is utterly impossible to live at all. Or, to express my

> theme even more simply: *There is a degree of sleeplessness, of rumination, of historical sensibility, that injures and ultimately destroys all living things, whether a human being, a people, or a culture.* ("OUL" 89; emphasis in original)

Yet, on the other hand, without a history that serves life, an invested, purposive, or directed history, we would be less rich in our resources to identify the characteristic weaknesses of the contemporary world; more significant, we would be unable to produce the resources needed to make a future that is uncontained by the present. So, an overburdening by history produces a kind of contemporary paralysis and a mode of self-justification for *any* action, however servile and contemptuous; but being uninfluenced by history means living ineffectively and without evaluations, in the pure immediacy of a present unanchored in time, that is, dislocated from the implications and effects of one's present. Leaving behind the past thus also means losing access to the future.

This is Nietzsche's own evaluation of his training as a classicist, his own untimely debt to the historical: "It is only to the extent that I am a student of more ancient times—above all, of ancient Greece—that I, as a child of our time, have had such unfashionable [untimely] experiences. But I have to concede this much to myself as someone who by occupation is a classical philologist, for I have no idea what the significance of classical philology would be in our age, if not to have an unfashionable [untimely] effect—that is to work against the time and thereby have an effect upon it, hopefully for the benefit of a future time" ("OUL" 86–87).

What history gives us is the possibility of being *untimely*, of placing ourselves outside the constraints, the limitations and blinkers of the present. This is precisely what it means to write for a future that the present cannot recognize: to develop, to cultivate the untimely, the out-of-place and the out-of-step. This access to the out-of-step can come only from the past and a certain uncomfortableness, a dis-ease, in the present. The task is to make elements of this past live again, to be reenergized through their untimely or anachronistic recall in the present. The past is what gives us that difference, that tension with the present which can move us to a future in which the present can no longer recognize itself.

There is a place for the ahistorical, the pure present. It is that which characterizes the inorganic, which requires no memory. Where the ahistorical marks the happy and contented time of the animal and the child who has as yet no history to recall, and where the historical is what differentiates the human from the animal, too much history, too great an absorption in the past inhibits actions in the present by positing too many possibilities and diminishing the sense of urgency of activity in the present. We require a certain threshold of history—not too much or too little, neither absorption in the past nor immunity from it in the present. History should not only promote life: privileging mere survival, as we recall, is only an elevation of the mediocre. Above all, history should be cultivated as necessary for the "higher life," as the means by which a higher life can raise itself above its present, all-too-human existence, to produce an unforeseen, self-surpassing future.

For Nietzsche, what is needed is a history that strengthens rather than weakens, that enhances rather than diminishes, that is assimilable into the present as a mode of the subject's self-expansion, that enables and encourages the powers of self-formation. Just as forgetfulness is that which most suits the ruminating nature of the animal and characterizes the helplessness of infants and the very old, so only the higher being can tolerate, and consume, welcome and live through, history. It is a qualitative measure, not a quantitative relation: only the strong can assimilate and digest history, live through and beyond it to become something more than the totality of one's history. The capacity to absorb and utilize the past is a test of the strength of the living:

> [We] . . . have to know exactly how great the *shaping power* of a human being, a people, a culture is: by shaping power I mean that power to develop its own singular character out of itself, to shape and assimilate what is past and alien, to heal wounds, to replace what has been lost, to recreate broken forms out of itself alone . . . The stronger the roots of a human being's innermost nature, the more of the past he will assimilate or forcibly appropriate; and the most powerful, the most mighty nature would be characterized by the fact that there would be no limit at which its historical sensibility would have a stifling and harmful effect; it would appropriate and incorporate into itself all that is past, what is its own as well as what is alien, transforming it, as it were, into its own blood. Such a nature knows how to forget whatever does not subdue it; these things no longer exist. ("OUL" 89–90)

The question is not *whether* to remember or to forget, but *what* to remember and to forget, when and in what context. A moderated diet of memory is needed, memory that does not overwhelm perception of the present and the edgy anticipation of the future, yet memory that does not let down the lure of the untimely, that facilitates the out-of-time of invention and innovation. The higher man, the future man, forgets that which he has accomplished, that which he has succeeded in and overcome: he is without ressentiment, content to take what he needs without guilt or memorialization, yet he must also remember what he has promised (for the promise propels the promiser into the future that is bound or somehow contained in the promise) and what he must face again, the consequences of his promises, the memory of his past action. The higher man digests what is of use of the past and forgets—expels—the rest, the past that he has absorbed, making him stronger, and the past he forgets, which enables him to retain his strength rather than dissipate it in rumination.

This may provide a possible path, with careful reworking, for any politics of the oppressed: to remember what one needs to move on to the future with more resources than the present can provide, while at the same time forgetting what one must, what has hurt, damaged, injured, or rendered one passive. A paradoxical impulse: to know what one must forget, to know in order to forget it. To make something positive of this past without betraying it, without repeating or continuing it, to produce a future that both breaks with the past yet at the same time refuses to disown it—this seems to be the very condition of all radical politics, from feminist politics that seeks redress for the harms done to women through the production of alternative histories, methods, and institutions; to postcolonial struggles which must address the degree to which the status of post- remains linked to the colonial; to the struggle of indigenous peoples to remain linked to their traditional beliefs and customs while at the same time attaining a viable place within an increasingly globalized economic landscape; and to holocaust survivors in their claims to recognition and reparation for past injury. In other words, at stake in this Nietzschean problem of the force of history is the degree to which the status of resistance (of the present/the future) is linked to, revives and transforms, the power (of the past).

Dynamic History

Nietzsche distinguishes among three types of history, three modes of reflection on the past: monumental, antiquarian, and critical history, whose calculated mix provides the ingredients for a healthy, vigorous capacity to live in the present and to seek out the force of the future. Each has its own functions and uses as well as its characteristic misuses. Each marks and relates to the human being in its own way. Monumental history is that which "pertains to him who acts and strives"; antiquarian history relates to the "one who preserves and venerates"; and critical history to the "one who suffers and is in need of liberation" ("OUL" 96). The higher man, the one who has no teachers or role models, who produces his own ideas and goals, must turn to history, indeed, has only history to turn to, to learn from, as the present can provide no clear clues to its own transformation. The history he acquires and learns from involves the admixture of these different modalities and functions of history, the recognition of the proper and improper uses of these histories, their strategic value in particular contexts. This is not the advocacy of a superior individual—or rather, it cannot be read in individual terms alone—for it is the characteristic of all politics, of all innovation and invention. How to create without reproducing prevailing models and practices, except to take what has been unutilized, undeveloped, from the past and make it a force that breaks through the strictures and limits on the present?

Monumental history is composed of the images and stories of individual great men and noble singular deeds, those that stand out from the mediocrity and pettiness of everyday life, to form a continuous chain of great and noble events leading to the present. In terms of feminist historical research, we have parallels in research that unearths or makes public the brave, great, and brilliant actions of individual women, usually forgotten in traditional forms of history. What is significant about such a history, what its utility might consist in, is both the premonition and anticipation of Nietzsche's account of the eternal return (that which is great is destined to return again), as well as the production of an ennobling lineage of ancestors, which may help call forth, produce, and refine the higher man, the "new" woman, the "post"colonial subject, the free individual, all subjects of their dynamic history of heroes and heroines: "That the great moments in the struggle of individuals form links in one single chain; that they combine to form a mountain range of humankind through the millennia; that for me the highest point of such a long-since-past moment is still alive, bright and great—this is the fundamental thought in the belief in humanity that expresses itself in the demand for a *monumental* history" ("OUL" 97).

What is of value in monumental history is not the production of truth, the creation of a genuine and direct lineage of rebellion, overcoming, uprising, but rather the production of a genealogy of greatness, of outstanding individuals, men or women, heroes, which may inspire acts of greatness in the present. But this very usefulness also prefigures the disadvantages, the liabilities for life of monumental history. For the great themselves, the monumental history of greatness has its own particular traps: "Men of power and achievement, deceived by beautiful images of triumph and mastery torn loose from the actual dense web of cross-cutting causes and effects in which all deeds occur, may be tempted to reckless, disastrous adventures in rewriting the past and scripting the future" (Berkowitz 1995, 32).

The great readers of the past, those who strive to emulate the heroic accomplishments of the past, live out, enact the eternal return themselves, though perhaps without knowing it. They demonstrate that what was once great can be great again, that what empowered once may serve to do so again. As disturbing as the misuse of monumental history by the great who overidentify with historical figures of greatness at the expense of understanding the machinations that made them great is the misuse of monumental history by the weak, the unfit, and the inept, who make images from the past act like frozen idols to be worshipped and to be kept in the present, who use these images to prevent themselves from acting in the present, who invest such images with nostalgia but with no current force, binding them in lost glory instead of reinvigorating them through their own actions.

By contrast, antiquarian history serves life by teaching a respect and reverence for the past. It serves communities and nations as much as individuals by unifying individuals into groups (ethnic groups, social and cultural communities) connected by the bonds of a common past, especially a common suffering, and a common earth. Antiquarian history makes and reflects communities, not through the worship of individual heroes, but in the reverence for artifacts, relics, and remainders from the past. It reveres small objects, fragments from the past, memorabilia, as if they were one's own personal historical documents. At its best, antiquarian history exhibits a certain reverence for things that resonate with and retain a significant connection with one's past, one's land, one's heritage or people. The goal of the antiquarian is to read and reread these things for the signs of his own belonging: "At times [the antiquarian] even greets across the distance of darkening and confusing centuries the soul of his people as his own soul; the ability to empathize with things and divine their greater significance, to detect traces that are almost extinguished, to instinctively read correctly a past frequently overwritten, to quickly understand the palimpsests, indeed polypsests—these are his gifts and his virtues" ("OUL" 103).

Antiquarian history degenerates quite readily into antiquarianism because it has no real perspective on the present. It tends to value everything old equally, to romanticize the old for its own sake. This degeneration occurs whenever the past and the old remain unconnected with the vitality of the present. Its danger is that it remains unanimated by current values. It manifests itself in the passion for collecting, for acquiring series, reproducing and reenacting the past in its detailed everydayness: "Antiquarian history degenerates from the moment when the fresh life of the present no longer animates and inspires it. At this point, piety withers, the scholarly habit persists without it and revolves with self-satisfied egotism around its own axis. Then we view the repugnant spectacle of a blind mania to collect, of a restless gathering together of everything that once existed. The human being envelops himself in the smell of mustiness" ("OUL" 105).

At its best, antiquarian history collects, preserves, and catalogues that which already exists, rather than generating something that does not yet exist. It lacks an impetus toward creation. It remains mired in attaining the things of the past rather than in harnessing the possible actions that animated their production for something new or renewed. This reverence for the past reduces concern for the present and commitment to the future. The new has no place in this schema, which is occupied only with preservation, which, in other words, remains stuck in memory without the necessary distance of forgetfulness.

To counteract the excesses of antiquarianism, a critical history is necessary, which is to say, a history that doesn't simply revere the past and memorialize its greatest figures, objects, and events, but that distances and dissolves the forces of the past in order to enable their reconfiguration in action in the present. It criticizes the past, and in this way functions as a self-consciously ethical or evaluative history, the only possibility thus far of a political history, a history of the forces that animate events whose implications continue into the present. We may today understand it as a political history, a history of the cost of events, deeds, and individuals (feminist or postcolonial history, for example), which, in criticizing past injustices and forms of suffering, overcomes them and transforms their significance and impact. This is the danger of critical history, but also its force:

> At times this very life that requires forgetfulness demands the temporary suspension of forgetfulness; this is when it is supposed to become absolutely clear precisely how unjust the existence of certain things—for example, a privilege, a caste or a dynasty—really is, and how much these things deserve to be destroyed. Its past is viewed critically, when we take a knife to its roots, when we cruelly trample on all forms of piety. It is always a dangerous process, one that is, in fact, dangerous for life itself; and human beings or ages that serve life by passing judgment on and

> destroying a past are always dangerous and endangered human beings and ages. For since we are, after all, the products of earlier generations, we are also the products of their aberrations, passions and errors—indeed, of their crimes; it is impossible to free ourselves completely from this chain. ("OUL" 107)

In other words, critical history reveals the violence and viciousness of human actions, the horror underlying all good deeds, the necessary underside of or accompaniment to life itself. Its danger is that what it makes explicit may serve to demoralize and prevent action, by revealing that man's "first nature," his historical condition, is steeped in horror and injustice. This first nature is that which produces patriarchy, racism, colonialism, slavery, whose heirs, in one form or another, are all of us in the present. This danger—remaining stuck in the past, seeing only its continuity and not its overcoming—can be averted only through the positive production of a "second nature" achieved only through "a new, stern discipline" that may replace this first nature. This is that self-overcoming required by the overman, man in his surpassed form: to accept as his own the history from which he comes, but to make this history the spur of a self-overcoming that severs him from that which produced him, to use history as a springboard for a future unmediated by and unnostalgic for the past and present.

These histories—monumental, antiquarian, and critical—may, when used in the right combination and context for the right purposes, serve the ends, not of mere life, but of ennobling life, of making life more and stronger than its past. A careful discernment of what we need from the past to remake ourselves in the future is what such histories, at their best, help provide. What may make history palatable is its active and vigorous use, its reverberations in the present. A healthy relationship between history and life involves the discernment of the life-enhancing features of the past and the leaving behind of what encumbers life in the present.

For Nietzsche, though, the problem is that the transformation of history into a scholarly discipline has meant that the connections with the past have been severed and need to be reestablished. The demand that history become a science renders it a mechanical process and mitigates the careful and calculated use of history by philosophy, and through art, by opening up the present to the massive weight of the past. Modern man, the man of learning in contemporary Germany, is a "walking encyclopedia," filled with facts, arcane bits of knowledge about the past, a passive recipient of received knowledges, and grateful to be reflecting and contemplating rather than acting. Because "history can be endured only by strong personalities; it completely extinguishes weak ones" ("OUL" 119), because the worship of the past veils the activity of the present unless it is adequately digested and excreted, history has tended to replace philosophy and art (production in concepts or images), or rather, history has infiltrated philosophy, it has replaced knowledge of and interest in this world with an obsession with the "facts" of the past: "Thus: history can be written only by the experienced and superior person. The person whose experience of some things is not greater and superior to the experience of all other people will also not be able to interpret the great and superior things in the past. The voice of the past is always the voice of an oracle; only if you are architects of the future and are familiar with the present will you understand the oracular voice of the past" (130).

A more adequate history would always rewrite the past, not only from the vantage point and interests of the present but, more significantly, from the interests of a future yet to be made, an audience yet to be born. It would revel in its own untimely location: "The genuine historian must have the power to recast what is age-old into something never heard before, to proclaim a general truth with such simplicity and profundity that we overlook the simplicity due to the profundity and the profundity due to the simplicity" ("OUL" 129).

There needs to be a tempered use of history, which in no way implies a mediocre, muted, or inaccurate history; it entails a history marked by its scholarship and accuracy. Such a history, above all, must be life-affirming, based on a love of or passion for life itself, which is to say, for the future and not just the past. Indeed, Nietzsche believes that fabrications or lies about history are justified to the degree that they foster and produce an active life. Yet history must remain rooted in truth in two senses. First, it must rest on the clear recognition of history before its concealment through the life-enhancing forces of the present: it must understand the deadly but true ("doctrines I hold to be deadly but true"): "the doctrines of sovereign becoming, of the fluidity of all concepts, types, and species, of the lack of any cardinal difference between human and animal" ("OUL" 153). Second, it must respond to and be rooted in "the real needs" of humans. Nietzsche's genuine historian aspires to truth and transforms history into art in order to educate and cultivate the noble impulses, to reanimate the history of the past by immersing it in the life of the future: "Only in love, only in the shadow of the illusion of love, does the human being create—that is, only in the unconditional belief in perfection and justness. Everyone who is forced no longer to love unconditionally has been cut off from the roots of his strength; he cannot help but wither, that is, become dishonest. In such effects art is the antithesis of history, and only when history allows itself to be transformed into a work of art, into a pure aesthetic structure, can it perhaps retain or even arouse instincts" (132).

As antidotes to "the historical sickness" that besets modern culture, subjectivity, and thought, Nietzsche advocates what he calls "the ahistorical and the suprahistorical." The ahistorical, not surprisingly, is to be identified with that power of forgetting that is so central to the stability and well-being of all life, and the suprahistorical must be understood as that which transcends history, or rather, marks all of history, is cross- or transhistorical. The ahistorical, as "the power to be able to *forget* and to enclose oneself in a limited *horizon*" ("OUL" 163), is that capacity to foreground, to repel the overwhelming onslaught of what is transient, temporary, provisional in order to focus on the present. But the suprahistorical marks out that which remains unchanged, that which survives and becomes what it is. These two antidotes move beyond the givenness of history and the inertia of the past to provide a way of dynamizing the past and making what is of value in it live on actively, not only into the present, but, in ways beyond our current control, into the future.

In his understanding of the value and limits of history, Nietzsche claimed that the past can be understood as triply enfolded, bound up with the dynamic movement, the force, of time itself. For the past is not merely a depleted resource, one robbed of its force or will, but is dynamic insofar as it remains the condition of the present surpassing itself. The past is:

1. The necessary condition for the present.
2. That through which the present has the resources to transform itself.
3. That which must be moved beyond and, if necessary, forgotten.

These claims are all closely linked to his more general understanding of time, particularly of the link between the past and the future, in his understanding of the eternal return, to which we turn in the next chapter.

The Will to Power

The will to power is, for Nietzsche, the only drive, the very drive governing not only all organic life, but all of matter as well. It is that which conserves the victorious forces of the past and directs them openly to the future. The will to power is the concept that Nietzsche offers Darwin to replace his account of natural selection, for the will to power is a positive active play of forces that produce and open up both matter and life. This concept, Nietzsche believed, overcomes what he saw as Darwin's muted and passive rendering of individual variations. It is his unique contribution to biology as well as to cosmology.

The will to power is a (nonpsychical, impersonal) will or impetus to more, to the increase of power, to the enhancement, not of a self or its ability to survive, but of its own forces, its own activities. This will is not what we understand as willpower, the self-consciously directed orientation to an attainable goal or object. Willing is always plural, multiple: there is never a single will to power, a single force. As Nietzsche says, "Willing seems to me to be above all something *complicated*, something that is a unity only as a word" (BGE #19). There is no singular force, no force assured of dominance. Rather, forces in the plural function and struggle within a field of all others. The will to power can exist and function only where there is a multiplicity of forces, each with its own wills, in other words, only where there is a differential of forces, where forces are stronger or weaker, active or reactive, commanding or obeying. Only where forces can differentiate and maximize themselves can there be a will to power.[1] Will is the force that commands, but it cannot command the world, matter, subjects (which would imply their passivity, their own operations outside force, outside will), but only other forces, other wills. The will to power can exist only where there is a commanding will and an obeying will, where the obeying will can command itself to obey:

> How does this come about? Thus I asked myself what persuades the living creature to obey and to command and to practise obedience even in commanding? Listen now to my teaching, you wisest men! Test in earnest whether I have crept into the heart of life itself and down to the roots of its heart!
>
> Where I found a living creature, there I found will to power; and even in the will of the servant I found the will of the master. (1965, *Thus Spoke Zarathustra* [hereafter *Z*], II, "Of Self-Overcoming")

The will to power is before, under, and beyond any self, sense of agency, or subjective intention. It is not the desire of a subject directed to an object, but a force directed only to its own expansion, for which objects take on the function of enabling or opposing forces. The will to power wills its own expansion and profusion, not as the Spinozist *conatus*, which seeks to retain its integrity, to retain its being, but rather as a mode of *maximization* of its being, the untimely explosion of force that activates, renders active, the will to power:

> Is "will to power" a *kind* of "will" or identical with the concept "will"? Is it the same thing as desiring? or *commanding*? . . .
>
> My proposition is: that the will of psychology hitherto is an unjustified generalization, that this will *does not exist at all*, that instead of grasping the idea of the development of one definite will into many forms, one has eliminated the character of the will by subtracting its contents, its "whither?" (WP #692)

The will to power is *not* a desire for power. As Nietzsche insists, there isn't a single, generic, universal will to power but a multiplicity of competing wills, which align to form organs, individuals, and collectives, which make up inorganic and organic objects as well as the fields within which these objects are to be located, each of which competes for whatever resources it needs for self-expansion without regard for the perspectives of the multiplicity of other forces. The will to power may prove to be another name for the principles of emergence, for the chaotic, competing distribution of forces in systems as they reach a point beyond equilibrium to destabilize and convert themselves to a different mode of organization. Although the will to power is not singular and unified, an overarching or global will, but a field, or series of fields, of force of varying intensity, it is the principle that underlies the world itself, the most fundamental principle of ontology, the single principle, for Nietzsche, governing all of existence. The will to power is a principle that operates through us and that makes us its modes of relay: "A quantum of force is equivalent to a quantum of drive, will, effect—more, it is nothing other than precisely this very driving, willing, effecting, and only owing to the seduction of language (and of fundamental errors of reason that are petrified in it) which conceives and misconceives all effects as conditioned by something that causes effects, by a 'subject,' it can appear otherwise . . . The popular mind separates out the lightning from its flash and takes the latter for an action, for the operation of a subject called lightning . . . But there is no 'being' behind doing, effecting, becoming; 'the doer' is merely a fiction added to the deed—the deed is everything" (GM 1 #13).

As subjects, we are as evanescent as lightning, as indiscernible from our effects as lightning from its flash. We are exemplars of the will to power, which exhibits itself in the very struggle of each of our cells, tissues, and organs, for oxygen, for nutrients, for activity, for intensification. As stable, unified subjects, we are effects, consequences, of the underlying struggle of this teeming multiplicity of organic (and other) forces that play themselves out through us. In each of us, there are wills to command and wills to obey, supra- and supersubjective wills, that lie below or above our consciousness. Even consciousness itself is the consequence of this multiplicity of wills—wills to be loved, to be independent, to be strong, to be vulnerable, to succeed, to fail, and so on. Everything that is animal and animate, everything that is living or even material, can be seen as the consequence of the struggle of a multiplicity of wills to power, wills to command or obey within us all. This will to power takes the place of what Nietzsche understands as the Darwinian desire to live, the desire for preservation. What takes its place is a struggle for all that is of value, the very struggle to make values, the struggle not for life, but for a superior life, a life that makes itself art, a life in the process of making itself other. The will to power is what impels us to seek a more noble, a higher life, a life beyond ourselves, in other words, a life imbued with and as art, the art of self-production.

Yet the question remains: Why *will*? Why ascribe to this force, this relation between forces, pure abstract forces, a term that is so psychologically laden, so linked to subjectivity and intentionality? The will to power is not the same as anonymous force. Forces themselves, disorganized, often random, make up the world. The world is nothing but the alignment of various material forces. The will to power cannot be separated from these determinate forces, which can be defined by their quantities, qualities, and directions. It inheres in particular alignments of force without being identical to these alignments.[2]

In Deleuze's (1983) terms, the will to power is both what forms, engenders the struggle between forces, their differentiation, as well as what coheres them into a field within which their competitive endeavors are made possible.[3] It is the attribute of both the victorious force and the entire field of forces. The will to power is what enables forces to align or compete, as well as what incites and selects the victorious forces. The will to power, in other words, must be understood as the *internal complement* of the victorious force and the inner resolution of the entire complex of forces: "The *victorious* concept 'force,' by means of which our physicists have created God and the world, still needs to be *completed*; an *inner* will must be *ascribed* to it, which I designate as 'will to power' " (WP #619).

To suggest that the will to power is an internal activation of forces, the activation of forces from their own point of view, the acquisition of a point of view by impersonal forces, is to recognize the pre- and superhuman resonance of this concept. The will to power has no will in any recognizable form: it is not the anthropomorphization and projection onto forces of human characteristics. Rather, wills to power compose human will, which itself teems with conflicting forces and renders the psychological itself inhuman: "In the chemical world the sharpest perception of the difference between forces reigns. But a protoplasm, which is a multiplicity of chemical forces, has only a vague and uncertain perception of a strange reality . . . To admit that there are perceptions in the inorganic world, and perceptions of an absolute exactitude; it is here that truth reigns! With the organic world imprecision and appearance begin" (Nietzsche, quoted in Deleuze 1983, 204).

The will to power becomes the internal determination of the strongest of forces, the "perceptual" facility a force can develop, its perspective or interest. The more simple the organism, and thus the organized collection of forces that compose it, the clearer and more precise its interests. These perspectives, interests, and perceptions are not limited to the animate, but exist everywhere. (In a sense, this is Nietzsche's untimely response to Dawkins: the gene certainly is selfish, self-directed, but so equally are the cell, the organ, the organism, the variety, the species, and even the environment; each has its own interests, and no one of these interests has an a priori privilege over any other. The gene is no more selfish than any other element constituting life!) It is for this reason that there are multiple wills to power, multiple interests and relations between the wills, interests that are themselves a ramification of their own will to power. The will to power is the inhuman, prehuman force that constitutes life and the human out of the chaotic organization of natural forces. Life and matter are nothing but the different alignments and coordinations of various wills to power.

If the will to power can be regarded as the acquisition by victorious forces of an internal will, a desire, an orientation to be more, then it must also be seen as that which underlies and makes possible the very field within which some forces come to function as dominant; as Deleuze describes it, it is the *synthesis* of forces, forces coordinated together in their relations of antagonism or cooperation. The will to power must be regarded as the inner affirmation of both the victorious forces and the collection, the synthesis of all forces. The power of synthesis that the will to power exerts over the collection or multiplicity of competing forces can be understood precisely as the eternal return, as the idea that all forces will recur again. (We shall return to this question in more detail in the next chapter.)

Instead of conceptualizing the subject as an agent of causal effects or a victim of another's agency, that is, as an intentionality, a will, a set of desires, especially as a "radical will" that acts and produces events, effects, that can be seen to conflict with the forces of social regulation, that is, instead of seeing politics as the more or less violent negotiation between individuals, groups, and institutions—individual and collective agents—Nietzsche may help provide a way of understanding politics, subjectivity, and the social as the consequence of the play of the multiplicity of impersonal active forces that have no agency, or are all that agency consists in. Which is to say, force needs to be understood in its full subhuman and superhuman resonances: as *the inhuman* that makes the human possible and at the same time positions the human in a world where force works in spite of and around the human, within and as the human. Politics may be understood as the attempt of the human to contain and direct an inhuman according to some prevailing interest—but the inhuman retains its force even as the human attempts to make it over into a human extension, an appliance or tool. The inhuman, forces functioning in their own interest, adds unseen or uncontrolled forces and effects, its own "interpretations" to the very (human) forces that attempt their control.

The World as Will to Power

In *Beyond Good and Evil*, Nietzsche engages in an "experiment" in which he wagers that the will to power may serve to problematize and complexify a whole series of terms usually related through opposition, providing linkages between terms commonly severed. Among these terms: real and apparent, mental and material, inner and outer, subject and object. The will to power, as the mobilization of forces, underlines and explains all these oppositional pairs. What if we imagine, he asks, that the entire universe is composed of the will to power? Can we explain relations between subjects and objects, mental and material entities, using this hypothesis? This conceptual experiment requires of us the following:

> Not to assume several kinds of causality until the experiment of making do with a single one has been pushed to its utmost limit (to the point of nonsense, if I may say so) . . . The question is in the end whether we really recognize the will as *efficient*, whether we believe in the causality of the will: if we do . . . then we have to make the experiment of positing the causality of the will hypothetically as the only one. "Will" of course can affect only "will"—and not "matter" (not "nerves" for example). In short, one has to risk the hypothesis whether will does not affect will wherever "effects" are

> recognized—and whether all mechanical occurrences are not, insofar as force is active in them, force of will, effects of will.
>
> Suppose, finally that we succeeded in explaining our entire instinctive life as the development and ramification of *one* basic form of the will—namely, the will to power, as *my* proposition has it; suppose all organic functions could be traced back to this will to power . . . The world viewed from the inside, the world defined and determined according to its "intelligible character"—it would be "will to power" and nothing else. (*BGE* #36)

This is an extraordinarily rich and fertile thought experiment which poses a crucial question, one of the questions of a basic ontology: Can we think the world in its teeming complexity, in its capacity to bring contraries into connection, in its vast and unexpected range and scope for encounters, through a single elementary principle? Not necessarily as a broad unifying principle, a law, that regulates all cases, but as the most elementary and incontrovertible principle that is relevant to, underlies, and is expressed in all cases? Could this single principle be the will to power? Can the will to power function as a principle that underlies, even if it does not unify, all that is mind, all that is matter, all that is inner, and all that is outer? In Nietzsche's formulation of this experiment, there are several salient features:

1. We can imagine the will to power as the set or interplay of forces that regulates the whole of the material world. The material world is nothing but the interplay of differential forces, forces of greater or weaker strength, gravitational, energetic, thermodynamic, electrical, magnetic, and so on, each of which strives to be more. These forces run indiscriminately through all relations, independent of their substance, in a certain sense leveling them, enabling them to interact. The will to power cannot be identified with these forces; rather, it must be understood as the alignment or coagulation of differential forces, forces in the plural, forces relative to each other. It is expressed by these forces and constitutes their totality, their alignments, and various tensions.

2. The biological world emerges out of the interplay of the forces of the material world, the playing out of its various wills to power. It is, in Nietzsche's understanding, a pre-form of life. The biological can be seen as a complexification of the forces of the material world, not different in kind, not directed to any transcendence, but of this one and only material world, a world that has its own differentiations. The organic world is necessarily connected to the inorganic world through the consumption and use of its inorganic forces: life sustains itself only through its capacity to extract from

and partake in the forces of the universe. Life extends and protracts force, holds on to it, grows through it; matter simply expends itself without protraction: "The connection between the inorganic and the organic must lie in the repelling force exercised by every atom of force. 'Life' would be defined as an enduring form of processes of the establishment of force, in which the different contenders grow unequally. To what extent resistance is present even in obedience; individual power is by no means surrendered. In the same way, there is in commanding an admission that the absolute power of the opponent has not been vanquished, incorporated, disintegrated. 'Obedience' and 'commanding' are forms of struggle" (WP #642).

3. All causality must be regarded, not as the interplay of discrete atoms or pulses of force, separate and discernable, but as the continuous efficacy of the will to power. Causation is our *feeling* of power resulting from an act of will: "If we translate the concept 'cause' back to the only sphere known to us, from which we have derived it, we cannot imagine any change that does not involve a will to power. We do not know how to explain a change except as the encroachment of one power upon another power" (WP #69). Will works only on will. To the extent that will works on matter, matter itself must be nothing but will. Cause is our way of retrospectively affirming the continuity of will through its effects: "Against the doctrine of the influence of the milieu and external causes [i.e., his reading of a Lamarckianism]: the force within is infinitely superior; much that looks like external influence is merely its adaptation from within. The very same milieus can be interpreted and exploited in opposite ways: there are no facts.—A genius is not explained in terms of such conditions of his origin" (#70).

The will to power, the affirmation and valorization of the differential play of forces, runs through all "things." Indeed, things themselves are nothing but perspectives on or "interpretations" of these fluctuating wills to power. The will to power interprets; above all, it evaluates, it provides the means by which evaluations are possible, the means by which evaluations enact themselves:

> One may not ask: "who then interprets?" for the interpretation itself is a form of the will to power, exists (but not as a "being" but as a process, a becoming) as an affect.
>
> The origin of "things" is wholly the work of that which imagines, thinks, wills, feels. The concept "thing" itself just as much as all qualities.—Even the "subject" is such a created entity, a "thing" like all others; a simplification with the object of defining the force which posits, invents,

thinks, as distinct from all individual positing, inventing, thinking as such. Thus, a capacity as distinct from all that is individual—fundamentally, action collectively considered with respect to all anticipated actions (actions and the probability of similar actions). (WP #556)

The will to power is that which differentiates itself, that which grows and strives for more, that which coagulates into entities and things, but which is itself the ceaseless struggle of differential forces, the struggle between different qualities and quantities of force. This is not the struggle for existence or preservation, as Darwin suggested, not the drive for pleasure and the avoidance of unpleasure, as Freud will claim, nor the stabilization of an identity or system, though each, when it occurs, is a consequence of the will to power. It is the play of forces, as well as the functioning of those forces that prevail:

> [My theory would be] that the will to power is the primitive form of affect, that all other affects are only developments of it;
>
> that it is notably enlightening to posit *power* in place of individual "happiness" (after which every living thing is supposed to be striving): "there is a striving for power, for an increase of power,"—pleasure is only a symptom of the feeling of power attained, a consciousness of a difference (—there is no striving for pleasure: but pleasure supervenes when that which is being striven for is attained: pleasure is an accompaniment, pleasure is not the motive—);
>
> that all driving force is will to power, that there is no other physical, dynamic or psychic force except this. (WP #688)

What distinguishes life from the inanimate forces of matter is not the will to power itself, for this will to power, in Nietzsche's hypothesis, is everywhere, suffusing all relations, its differentiated modes of expending and retaining force. Life is that which accumulates force, attracts other wills, augments itself. It refuses or surpasses, at least temporarily, the first law of thermodynamics (the principle of the conservation of energy), for life is that which always functions in excess of need, survival, stability, and pleasure. It sustains itself as a living being by ensuring that the expenditure of energy that life entails is matched by the accumulation of energy in consumption. Yet life is not balance and equilibrium but accumulation or expenditure to excess, the production of the unnecessary, invention and art as well as brutality and cruelty, for its own sake. Life is the supreme value, above identity, above being, above the human, and above knowledge. It is life itself that accumulates force and will:

> The will to accumulate force is special to the phenomena of life, to nourishment, procreation, inheritance—to society, state, custom, authority. Should we not be permitted to assume this will as a motive cause in chemistry too?—and in the cosmic order?
>
> Not merely the conservation of energy, but maximal economy in use, so the only reality is the will to grow stronger of every center of force—not self-preservation, but the will to appropriate, dominate, increase, grow stronger. (*WP* #689)

Nietzsche's provocation to philosophy is his contention that the will to power is the underlying explanation of the world, of those who live in that world, and of philosophy itself: "This world is the will to power—and nothing besides! And you yourselves are also this will to power—and nothing besides!" (*WP* #1067). The moral question that emerges out of such an ontology is this: If we are of this world, made up of the same forces as those which compose this one and only this world, how can we live this inhuman force that makes us? How can we affirm, instead of the attributes of a subject (intention, identity, recognition, autonomy, pleasure or desire), the primacy of the will, the prevalence of dominant forces, the impersonal that structures the world of persons? How, in other words, can we affirm the eternal return which both signals our demise and surpassing as subjects, human beings, and life forms and promises a becoming beyond the human, a future beyond recognition? For this, we need to understand the eternal return, the object of investigation of the next chapter.

6. THE ETERNAL RETURN AND THE OVERMAN

In *Thus Spoke Zarathustra*, Nietzsche claims, through the mouth of Zarathustra the prophet, teacher of the eternal return, that we need some kind of remedy for the allure and fixations that the past holds for us, the debilitations and the impotence it leads us to feel when we recognize that we are helpless to change it, the past that weighs on us so heavily. The will itself is overpowered by the burden of the past, which strikes a dark shadow over the future. Zarathustra seeks a doctrine that will liberate the will from its nostalgia for the past and that will orient it away from the past from which it sprang to the future it will produce. That doctrine, which Zarathustra believes will restore to man the "innocence of becoming," which will enable man to shake off the lure of the past by showing that the past itself provides no justification or value for the present, is the doctrine of the eternal return.

Time as Eternal Return

The will to power is central to Nietzsche's concept of the eternal return. The eternal return is what Nietzsche believes is necessary to revitalize the Darwinian notion of evolutionary development by positing, not an origin but a goal, a direction, for becoming: only the overman, only that which is beyond man and beyond life as we know it, only what is strong enough to face the future, a future inevitably bound to repetition, only a being who sees its strength as the surpassing of its past, a being fully oriented to the force of time itself, can provide a goal or telos for evolution. Nietzsche gives a value to Darwinism which, as a scientific doctrine, it has from its inception attempted to elide: not the value of privileging one of its component parts (the gene, the individual, group, or species), the parts that make up its processes, but value for the process itself, which produces, through chance, life with its inevitable push toward the future, an irreducible direction forward, eternal becoming. The hypothesis of the eternal return distinguishes Nietzsche's position not only from Darwinian biology, but also from Newtonian, Cartesian, and Einsteinian cosmologies, which configure time as a countable line, one of the axes that also represent space. It also challenges the contemporary fixation on science as the discovery of more and more minute and intricate forms of determinism, that is, on those models that reduce the future to the past, where current causes will give rise to predictable future effects. The eternal return is Nietzsche's philosophical, ontological provocation to science's atomistic, mathematical focus.

In spite of his attempts to distance himself from the science of his period, Nietzsche also remained intrigued by its ontological implications. His hypothesis of eternal recurrence links his conceptualization to nineteenth-century thermodynamics, which he uses as the latent framework or skeleton for his account;[1] his connection to the Darwinian open-endedness of evolution is also clearly stronger than his own statements acknowledge. In many ways, the eternal return can be seen as a curious amalgam and a bizarre, twisted reformulation of Darwinism, thermodynamics, and Kantian ethics.

His postulate of the eternal return is, arguably, one of the most difficult and misunderstood areas in Nietzsche studies: the status and place of time and of the eternal return is the most elusive, the most readily disposable element of Nietzsche's work, the one that many commentators find the least defensible of his numerous difficult-to-defend conjectures. Nevertheless, the eternal return can be read as the very center of Nietzsche's philosophy, his most unpalatable and intriguing hypothesis.[2] It needs to be understood in two quite different, yet related senses, which dangle under each other, never clearly separable, oriented toward two different goals and purposes. The eternal return is simultaneously the highest spur and moral principle of his transvaluation of values, both the means and the end of all his writings, and the most physical and metaphysical, the most ontologically committed principle of material existence in his speculations, the one most solidly connected to his conception of the way the universe is.

What is the relation between the will to power and the eternal return? To answer this question we must turn to a more preliminary question: What is the eternal return? What is it that returns? The will to power, as we recall from the previous chapter, is the active inner principle of both victorious forces and the synthesis of forces as a field of disparate intensities. The eternal return is the regulative cosmological principle of the material world (it is the regularity of the synthesis of material relations) which the will to power must, in spite of itself, adopt as its own in order to affirm and thus produce its maximal intensification. It is almost as if Nietzsche understood this as a relation of hierarchical elaboration or emergence, whereby the more chaotic level, field, or plane of competing disorganized forces finds its inner dynamic principle in the will to power. Each level or plane must take on the qualities of the one below it, as well as exceed those qualities, to emerge as a new and higher-order complexity. The will to power affirms force in its generality—the force that generates fields, planes, that coheres and enables forces to compete, particularly the victorious forces, active forces, which take themselves as far as they can go, which are not separated from their powers of action. This will to power, the inner accompaniment of the victorious forces, and the expression of the entire field of forces, finds its highest affirmation in the eternal return, in the commitment to the infinite repeatability of will, its ability to bear the scrutiny of infinite repetition, its strength to will itself to eternity.

Nietzsche claims in *Ecce Homo* (1967 [hereafter *EH*]) that the eternal return is his greatest accomplishment, the one to which he devoted himself wholeheartedly in *Thus Spoke Zarathustra*, a book that celebrates the vision, or rather, the mere insinuation, of the eternal return. Nietzsche describes the notion of eternal recurrence as Zarathustra's most "fundamental conception." He argues in a number of fragments that it is the core of his prescriptions for life and health. He considers it a bridge between his diagnosis of contemporary life and the possibilities for health, between the past and the future, a kind of welcoming of a future in which the eternal return not only adjudicates who, which one, is strong enough to overcome himself, but also establishes the practice of affirmation itself as part of this overcoming. The eternal return is the path between the all-too-human and the overman, between the contingency of the will to power or active force and the necessity of their evolutionary overcoming or becoming-other.

The Gay Science provides the earliest introduction of the notion of the eternal return. It already closely approximates his more mature understanding of the doctrine. He regards this idea as simultaneously the most terrifying and exhilarating of all concepts, and in this sense it is the greatest weight, the burden life must bear in order to overcome itself. In this well-known passage, Nietzsche envisions a demon revealing this most terrible and awesome thought:

> *The greatest weight.*—What if, some day or night, a demon were to steal after you into your loneliest hour and say to you: "This life as you now live it and have lived it, you will have to live once more and innumerable times more; and there will be nothing new in it, but every pain and every joy and every thought and sigh and everything unutterably small or great in your life will have to return to you, all in the same succession and sequence . . ."
>
> Would you not throw yourself down and gnash your teeth and curse the demon who spoke thus? Or have you once experienced a tremendous moment when you would have answered him: "You are a god and never have I heard anything more divine" . . . ? (GS #341)

The eternal return—the idea that everything that once was will recur over and over—is both the "greatest weight," an unbearable burden, an impossible idea, and the most invigorating promise, the very heart of the affirmation of will, of chance, of the moment.[3] The thought of the eternal return is intolerable insofar as it implies living every frustration, humiliation, imperfection, every detail, every lie and failing, over again. Its weightiness, its burden is readily clear. What is less clear is the source of its joyfulness, its triumphal affirmation, its positivity, its capacity to invigorate, to make whole. "To the paralyzing sense of general disintegration and incompleteness I oppose the *eternal recurrence*" (WP #417). It is an ethical doctrine insofar as it functions as a test of the will, a test of its moral strength: Will this action such that it will eternally recur, for it will in any case recur again. And it is the criterion of value as well: affirm necessity; make the highest virtue from what one must by making it one's own.

Inseparable from its moral weight as both test and goal of the will to power, the eternal return is also a fundamental ontological principle, the most fundamental principle regulating the universe in its entirety, regulating all forces in all of their possible combinations: cosmological, geological, historical, cultural, and psychological. It is the most basic postulate on which all others are developed, an ontology as powerful—if not as effective or scientifically validated—as Darwin's own. It is the culmination of Nietzsche's understanding of a world populated by teeming, proliferating, competing wills to power. The eternal return becomes the formative doctrine of Nietzschean physics and, through it, Nietzschean biology. It is only as ontological principle that the eternal return can come to have its force of moral imperative: to accept and affirm with carefree joy the necessity of will and the necessity that what the will wills once, it will will again and again.

Eternal Return as Cosmological Principle

The eternal return thus has two dimensions. The first is as a physical and ontological doctrine; in this sense, Nietzsche links the principle of the eternal return to nineteenth-century cosmology and thermodynamics. The second is as an ethical and transvaluative process, a mode of impossible ideal, a moral test by which only the strongest can produce a transvaluation of existing values. To look in more detail, first, at the ontological principle: Nietzsche regards his doctrine as an alternative to, possibly even if it is an indirect, wayward implication of, the first and second laws of thermodynamics, the law of the conservation of energy and the principle of the maximization of entropy, respectively, which, taken together, imply that the universe as a whole is gradually winding down, heading toward "heat death": "The law of the conservation of energy demands *eternal recurrence*" (WP #1063).

Against the postulate of a beginning and an end to the universe—the driving concern of much of contemporary cosmology—Nietzsche hypothesizes the infinity of time, coupled with a finite series of things, states, and events, a finite amount of matter and energy, which together produce a machinery for the generation of an extremely large but nevertheless finite combination of states and events, and thus the inevitability of their eternal occurrence. Time's vast, overwhelming infinity selects every variant, every combination, sooner or later, given a long enough span:

> If the world may be thought of as a certain definite quantity of force and as a certain definite number of centers of force—and every other representation remains indefinite and therefore useless—it follows that, in the great dice game of existence, it must pass through a calculable number of combinations. In infinite time, every possible combination would at some time or another be realized; more: it would be realized an infinite number of times. And since between every combination and its next recurrence all other possible combinations would have to take place, and each of these combinations conditions the entire sequence of combinations in the same series, a circular movement of absolutely identical series is thus demonstrated: the world as a circular movement that has already repeated itself infinitely often and plays its game in infinitum. (WP #1066)

A number of Nietzsche's commentators have puzzled over these speculations. For example, Linda Williams (2001, 112), in her recent study, claims that the cosmological doctrine has three fatal problems: it assumes the finiteness of both space and matter, an assumption that is no more warranted than the postulate of an infinity of time; it cannot adequately distinguish between a sequence and the repetition of a sequence, a repetition that must include within it the trace of the earlier sequence and thus cannot be conceived as an exact repetition; and, even if it were true as a doctrine regarding the universe as a whole, the time scale that it represents makes it irrelevant to our daily lives, or even to the time of humanity. David Wood (2001, 16) agrees that Nietzsche's argument is "not convincing" and, even if true, is irrelevant to us. Like Williams, he suggests that we need a principle, which Nietzsche himself does not provide, to index the number of repetitions, to distinguish one repetition from another, which, ex hypothesi, is not possible, for each repetition must be identical.

Danto (1968, 205) also argues that "this is exceedingly garbled," claiming that Nietzsche awkwardly brings together a theory of probability (the dice throw) with a theory of the necessity of the repetition of exact combinations. It is not surprising that he finds the account perplexing. What he misunderstands is that Nietzsche is not suggesting the exact repetition of every event for every throw of the dice, but merely that sooner or later, everything will recur again. Every combination will inevitably occur, but not always and only with the precise details in which it once occurred. Nietzsche's account of the dice throw cannot be construed as an account of probability, an understanding of relative percentages, chances of possible outcomes over a statistically relevant number of throws (hedging one's bets). It is only an affirmation of will, as Deleuze makes clear, if the entire stakes are being wagered on a single dice throw. Probability is always, in fact, in the long run, necessity: what is probable—as well as what is improbable—will, given an infinity of time, happen, and happen again. These are not two accounts Nietzsche has cobbled together, one of probabilistic determinism, and the other of exact repetition; rather, there is a single account of the infinity of time and the finitude of matter, force, quantity, or energy, a single theory of a machinery of difference (the dice throw, the plethora of material forces) cast within the active dynamism of time, a single time in which matter and its events repeat and play out every combination, rather than a single script that matter infinitely replays.

The throw of the dice is one of Nietzsche's favorite images of the eternal return, one of the gestures of Zarathustra, whom Deleuze (1983, 88) describes as the "dark precursor" of the eternal return. In Deleuze's reading, the dice throw must represent pure chance insofar as its throw precludes the calculation made possible by several throws. A single throw of the dice, the double movement of falling up and down, initiates the play of chance alone. Repeating the gesture, figuring out probabilities through several throws reduces chance to predictability. The game is to throw a single time and affirm with all of one's being the chance that necessity falls to earth. Nevertheless, while affirming the pure element of chance, the throw of the dice also brings with it the structure of necessity: the dice that are thrown affirm chance; the dice falling back to earth affirm necessity. Necessity does not obliterate chance but affirms it, for necessity is not the opposite of chance, but its difference. The eternal return is Nietzsche's maximal affirmation of the chance of the dice throw: Throw such that you not only live joyously with the (random) consequences of that one event, but even more so, throw such that its consequence will be eternally repeated. The eternal return, exemplified by a single throw of the dice, thus both presupposes and brings repetition into existence; and it is the repetitions it engenders that generate and produce difference.[4]

What returns is not the identical, but only the extreme, the active, that which, in going as far as it can, brings about its own changing, changes into something else, metamorphoses itself. What returns, or at least, what never ceased and has no need of return, is the force of time, the differentiating forces that arrange and rearrange things and their states. Return is that which initiates, and engenders, difference! Deleuze proposes a double affirmation of Nietzsche's insight: the single throw of the dice is the affirmation of the being of becoming. The being of this becoming is precisely the realm of the virtual, the universe of intensities, of the past, of problems/ideas, the realm Bergson describes as consciousness; and the becoming of the being of the world itself, which is that of difference, of the becoming or actualization of extensities: "The eternal return is neither qualitative nor extensive but intensive, purely intensive. In other words, it is said of difference. This is the fundamental connection between the eternal return and the will to power. The one does not hold without the other. The will to power is the flashing world of metamorphoses, of communicating intensities, differences of differences, of *breaths*, insinuations or exhalations . . . Eternal return is the being of this world, the only Same which is said of this world and excludes any prior identity therein" (Deleuze 1994, 243).

The will to power is the world of infinite change; the eternal return is the very being of the world. To affirm the being of becoming is to respond to the intensity of the provocation to thought and to affirm the virtual, that is to say, the as-yet-untapped impact of thought. The domain of the virtual is the domain of eternal return, of difference, of time itself.[5]

What Nietzsche affirms is the fundamental logical and ontological separation and independence of time from matter, though not of matter from time. Time must be understood as an infinity, stretching both backward into the past forever and ahead indefinitely into the future, a passage with no beginning or end. Thus, time itself cannot come into existence or cease existing; it cannot begin or cease to become. Insofar as time exists now, it is always becoming, never having begun nor being able to cease. The eternal return is the affirmation of the infinite becoming of time. It is not clear that this infinity of time is best represented through the infinite extension of the line; it is even less clear that Nietzsche's conception of time can be represented as cyclical. Nietzsche is not suggesting that time is linear; it is thus incapable of being construed as a circle. What recycles is never time itself but what exists in time: things, processes, events, formations, constellations, in short, matter in all its permutations.

It is only through this perpetual becoming of time that the very semblance of being is possible: "If the universe were capable of permanence and fixity, and if there were in its entire course a single moment of being in the strict sense it could no longer have anything to do with becoming, thus one could no longer think or observe any becoming whatever" (*WP* #1062).

With the exception of the great theorist of flux, Heraclitus, philosophy in general performs a double reduction: it reduces the becoming of time into the being of things, by reducing becoming to being and time to things. Yet the being of time is this very peculiar becoming which touches all things with its directed movement forward to the future while always looping relentlessly back over the past, always carrying and transforming the past as it transforms itself to give way to the future. Becoming cannot start or finish without ceasing to become. This is why, if there were even a single moment of being, all of becoming would freeze, would cease to become. This pure becoming that underlies the universe is what the eternal return affirms: it is the telos (which is not one) that Nietzsche provides for Darwinian becoming. It does not affirm that which returns, but the very process of returning itself: "That everything recurs is the closest approximation of a world of becoming to a world of being" (*WP* #617).

Deleuze (1983, 48) acknowledges that this is a key to understanding the force of the eternal return: "Returning is the being of that which becomes."[6] The affirmation of returning forever is the affirmation of the very being of becoming, of what it is that becomes: returning is thus the highest affirmation of the power of time itself, the "life" of time. The eternal return cannot be understood, as Danto reads it, as a doctrine of the repetition of all things, acts, and identities. It is not being, matter, things that return: rather, it is returning that is the being (i.e., the becoming) of things and of matter in its particular configurations, including those constituting life.[7]

Nietzsche's cosmological conception of the eternal return, however, may not be as far-fetched, as idiosyncratic, as it appears to some commentators. Although it is largely discussed in embarrassed tones and by common agreement regarded as a sign of Nietzsche's eccentricity, it does not today, in the light of the most speculative outreaches of contemporary cosmology, seem quite so peculiar or anachronistic. Perhaps Nietzsche once again anticipated an audience to come—not the audience of his own scientific contemporaries, but ours. Perhaps, it is implied, one of the greatest follies of our contemporary cosmology is precisely the scientific conflation, since Einstein, of time with matter (the equation or reduction of time to some variable of matter), particularly with the movement of light, which results in a philosophical conundrum. The question What was occurring before the Big Bang, that is, what happened before the birth of matter in the form of the newly born universe? has been regarded as incoherent and meaningless in the mainstream sciences. But Nietzsche asserts that it makes sense to ask the question What happened before? even if we cannot provide an adequate or satisfying answer. It makes sense because time and matter are conceptually separate, because the qualities of time are not the same as those of matter. Matter conserves itself in the universe as a whole, but time has an infinite span, it squanders itself without loss. It follows that in the infinity of time past and time future, every conceivable combination of matter has already occurred, and will occur again, an infinite number of times. This is the full force of the concept of eternity, of eternal becoming. Eternity is not stillness, the unchanging, the immutable, but endlessly varying difference, difference that ends up exploring every element of phase space, every possible combination, probable and improbable. Time's infinity can never be reduced to the movements of matter, for time explains matter's endless capacity to become-other:

> The total amount of energy is limited, not "infinite." Let us beware of such conceptual excesses! Consequently, the number of states, combinations, changes and evolutions of this energy is tremendously great and practically immeasurable, but in any case finite and not infinite. But the

> time through which this total energy works is infinite. That means the energy is forever the same and forever active. An infinity has already passed away before this present moment. That means that all possible developments must have taken place already. Consequently, the present development is a repetition, and thus also that which gave rise to it, and that which arises from it, and so backward and forward again! Insofar as the totality of states of energy always recurs, everything happens innumerable times. (Nietzsche, from *Werke,* quoted in Danto 1968, 205)

Time is unlimited, infinite, suggests Nietzsche: there is no scarcity or rarity of time, just as, for him, we do not exist in a condition of fundamental economic or energetic scarcity on earth, as Malthusianism implies, but in temporal and material abundance and excess. The time of the past is infinite, which means that becoming could never have started but must always be: if there were a final state of the universe, it would have already been reached, given this infinity of the past: "If the world could in any way become dry, dead, *nothing* or if it could reach a state of equilibrium or if it had any kind of goal that involved duration, immutability, the once-and-for-all (in short, speaking metaphysically: if becoming *could* resolve itself into being or into nothingness), then this state must have been reached. But it has not been reached" (WP #1066).[8]

The time of the universe is unlimited in both directions. But the matter of the universe, or equally, its energy, is limited, finite, and is blighted by the prospect of the gradual winding down and dissipation predicted by the second law of thermodynamics. The universe itself, science has told us, is gradually cooling down as it expands outward. Eventually, even planetary objects that live will die; the universe itself is subjected, over a massive time scale, to this gradual unwinding. It is this limited or finite amount of energy that contemporary physics focuses on in its various accounts of the origins of the universe, and it is through an extrapolated reversal of the movement of expansion that a beginning has been posited.

It is quite remarkable, given the power and dominance in physics of the theory of relativity, and in cosmology of Einstein's conception of an expanding universe whose speed dictates the forms space and time themselves take, that the current focus on reconciling Einstein's general theory of relativity with the more peculiar pronouncements of subatomic or quantum physics has resulted in the return to something like a Nietzschean understanding of the eternal return. Only very recently have scientists (such as Andrei Linde [1994], Alan Guth [Guth and Lightman 1998], and other proponents of the eternal inflation theory) argued that the Big Bang is not an origin, the birth of the universe, but a transition, a kind of quantum leap between one universe at its death and the birth of another. The Big Bang is not the only generative explosion of matter into an expanding universe; it is only the current explosion, having been preceded by infinite earlier Big Bangs to be followed by infinite later Big Bangs.[9]

If one posits the infinity of time and the finitude of matter, if one posits time as conceptually independent of matter, then Nietzsche's hypothesis seems inevitable rather than garbled: everything that will happen has already happened in just the way it is happening now. And in every other configuration. And it will happen again, infinitely. This is not to say that every other combination and possibility does not also occur; rather, given enough time, what has once occurred, what might or could occur, will inevitably occur again, will play itself out in exactly the same terms. Every possible combination will play itself out, given the infinity of time. There can be no Hegelian sublation, no overcoming of what has been superseded, no ridding ourselves of already worked out possibilities. The infinity of time, and its relentless activity in governing all material objects and processes, guarantees that the same combination, the same alignment of forces will occur again. Everything occurs again. But only what actively affirms itself affirms and celebrates returning itself, and can elevate itself through this acceptance of repetition, becoming, the impossibility of remaining the same. The celebration of the eternal return, the prerogative only of the strongest, is the affirmation of time, the revelry in life as an affirmation of the becoming of time, the double affirmation both of the particularity of each moment and also the force of the movement of all moments together: " 'Timelessness' to be rejected. At any precise moment of a force, the absolute conditionality of a new distribution of all its forces is given; it cannot stand still. 'Change' belongs to the essence, therefore also temporality: with this, however, the necessity of change has only been posited once more conceptually" (WP #1064).

Rather than an infinite repetition of exact identity or sameness—everything returns exactly the same, over and over, every detail reappears in perpetuity, in the exact order, with the same consequences—what returns is the *synthesis* of difference: "According to Nietzsche the eternal return is in no sense a thought of the identical but rather a thought of synthesis, a thought of the absolutely different which calls for a new principle outside science. This principle is that of the reproduction of diversity as such, of the repetition of difference" (Deleuze 1983, 46).

As cosmological principle, the eternal return can be understood as the affirmation of the synthetic and recapitulative force of time itself, the acknowledgment of the independence and autonomy of time and thus becoming from the arrangements of matter, the capacity of time to rewrite itself, its past, and to open out onto its future. The eternal return, while positing the inevitable and eventual return of what has already happened, nevertheless, perhaps paradoxically, does not preempt or close down the future. Rather, it welcomes the future as the culmination and reiteration of the present, as a synthetic overcoming of the present. It welcomes it in the manner of a wager that welcomes the outcome, win or lose, that confirms the value of the wager itself. The eternal return in no way renders the future more predictable; if anything, it ensures that the connections between present and future are not simply one-way, where the present causally prestructures future effects, but two-way, where, in spite of causal connections, other relations (iteration, coalescence, retrospection) also emerge without predictability to link the future, and the past, to the present.

Today, Nietzsche's physical doctrine of the eternal return appears in the light of positivist science to be an unconvincing argument, steeped in metaphysics, and one that, not surprisingly, flies in the face of current physics and cosmology (which, it must be admitted, are also the offspring of nineteenth-century mechanism, of which Nietzsche is so critical but on which he so heavily relies). It is surprising, however, that certain traces of Nietzscheanism, conceptually experimental pockets of the sciences with Nietzschean quirkiness, reappear in the hard sciences at precisely those times when positivism seems to consolidate itself most decisively. The enigmas of quantum mechanics and of an understanding of the universe in terms of the independence of time from space and from matter are still being explored. The implications of Heisenberg's uncertainty principle—that subatomic particles emerge into and out of existence, that they do not occupy a definite or determinable location or speed, that space can never really be conceptualized as empty, that quantum fields may provide the energy for particles to emerge into or disappear from a particular location, that an electron, for example, might be regarded as spread out through the entirety of space until it is measured or observed at some specific location, and that perhaps the universe itself is spread out over all space until it is observed and located—share certain resonances with Nietzsche's own, quite wild speculations about cosmology. Although physicists and cosmologists would like to create a regular and knowable universe, the more deeply and minutely they observe and explore its elements, the more regularly the universe resists simplistic explanations and complete predictability. There is a certain becoming-Nietzschean at the most speculative outcrops of current-day physics, even if the mainstream remains mired in the tones of a theological passion Nietzsche denounces, sure of its project, convinced of outcomes. Elsewhere, in the more perilous reaches of scientific speculation, there may be more of a return to Nietzsche than scientists would be happy to acknowledge.

Eternal Return as Moral Principle

More convincing, and currently more credible than the physical doctrine, though perhaps not as readily separable from it as many would hope, is the ethical and transvaluative dimension of Nietzsche's claims.[10] A number of scholars, including Nehemas (1985) and Deleuze (1983), have argued that it is not clear that Nietzsche requires the physical postulate to uphold the ethical postulate, and that they can function independent of one another, though I suggest that the connection is more intimate between the two doctrines than any easy separation would allow. Without its force as the most basic physical principle, the eternal return lacks the weighty gravity needed to enable it to function as moral principle. Whereas the ethical doctrine requires the ontological grounding of its moral force in the world itself, it is less clear that the ontological and physical doctrine requires the support of the moral principle. The ontological hypothesis retains its force whether or not there can be a moral imperative for living beings—indeed, whether or not there are living beings.

The eternal return must be regarded as the supreme value, the value that sorts through and settles the value of other values. It is only the strongest of wills that can will itself to eternity, can affirm its choices and actions forever. To live the eternal return, to revel in its implications, to accept them fully and generously, is to accede to a higher life, the highest life we can still recognize as human, that life which comes closest to approximating the overman, man's evolution beyond himself. The eternal return is the test of this height, for it requires, produces, or selects only the strongest of characters, only those who can endure its horrible, liberating prophesy: "To *endure* the idea of the recurrence one needs: freedom from morality; new means against the fact of *pain* (pain conceived as a tool, as the father of pleasure; there is no cumulative consciousness of displeasure); the enjoyment of all kinds of uncertainty, experimentalism, as a counterweight to this extreme fatalism; abolition of the concept of necessity; abolition of 'will'; abolition of 'knowledge-in-itself' " (WP #1060).

Only the strongest—that is, the bravest, the most singular, the least nostalgic, the least absorbed in the past, the least reliant on pregiven values, the most prepared to produce one's own values, to use obstacles as one's mode of action, to enjoy uncertainty, to give up will, desire, knowledge—can bear the weight of this comprehension, the burden of this necessity: to live again, to live everything again, to will now in the present to live everything again in the future. A curious and horrifying inversion of the Kantian categorical imperative: not the legislative universalization of one's action as a template for all to behave in the same manner, but the internal universalization of one's own actions for oneself forever. What Kant attempts to universalize as it were spatially, within living populations, Nietzsche universalizes temporally, for a singular life. Deleuze (1983, 68) transcribes the formulation of the eternal return in the language of Kant to indicate its debt to practical synthesis: "Whatever you will, will it in such a way that you also will its eternal return."

Nietzsche links the eternal return not only to the central concept of the will to power, but also directly to his concept of the overman, or *Übermensch*.[11] The overman is, ironically, a fundamentally Darwinian character, an evolutionary development from man: as the ape functions as an earlier progenitor of man, so man functions, ape-like, as an earlier progenitor of the overman. The overman is the highest possibility of man's evolution beyond man, beyond even the highest man, man's only hope of living on this earth into the future, beyond himself; he is the future becoming of man.[12] The overman is not the fittest man, exponentially developed out of man as we know him today, but the development of what is highest, noblest, strongest in man that can maximize itself only by a becoming beyond man. This overman is an evolutionary product, rising higher, as man does relative to the worm, to some indeterminate evolutionary height from which he can look back, amused, at that from which he came.

The affirmation of the eternal return is the prerogative only of the overman, for it could and does crush the frailty of the human, though it is only through such a destruction, overcoming, and forgetting that the overman is welcomed into existence. The overman is the one who, as a being of temporality, can, indeed must, produce that which does not decay and drop away, what is permanent, that which occupies the future, in art, knowledge, and morality, that which is not erased over time, not subjected to history, but which has its effects over and over again as eternity, in the suprahistorical. These are not a priori principles, universal laws, God-given imperatives imposed uniformly throughout time but becomings affirmed to their utmost as tenuous, provisional, deferring, to-be-elaborated, becomings that come again. Only the overman can survive the demand for a return; the overman is created by this very capacity and desire for return. The eternal return is more than the human can bare. The human and the living always return once, their phylogeny recapitulating ontogeny, their life repeating, through difference, the genetic codes the past has delivered to them, making of those codes something livable by being new. The overman, by contrast, can bear repetition not just once, as does the human, but an infinite number of times, can bear the idea that everything be infinitely repeated. The overman is the only "subject" of the will to power and the only goal of the eternal return.

So, what is it in the eternal return that returns? Is it everything, in its exact configuration, like the frame-by-frame replaying of a movie, as Danto (1968) and Berkowitz (1995) suggest? Is it the repetition of random fragments of a configuration, as contemporary genetics asserts? Or is it a selective or directed repetition, the repetition only of the victorious or the active, as Deleuze (1983) and Erich Heller (1988) argue?[13] What is it that repetition selects: everything, in its detailed combination; some things, those that are randomly preserved; or particular things, those that preserve themselves?

For Deleuze (1983), whose reading remains contentious among other commentators, the eternal return is that which, like natural selection, weeds out active from reactive forces: the very act of willing to eternity is what converts reactive back into active, for an act cannot simply be undertaken, it must be willed, and willing infinite repetition is itself an act of conversion. It transmutes reactive into active: "Laziness, stupidity, baseness, cowardice or spitefulness [the characteristic features of ressentiment] that would will its own eternal return would no longer be the same laziness, stupidity etc. How does the eternal return perform selection here? It is the *thought* of the eternal return that selects. It makes willing something whole. The thought of the eternal return eliminates from willing everything which falls outside the eternal return, it makes willing a creation, it brings about the equation 'willing = creation' " (69).

The very thought of the eternal return functions to negatively select, that is, to eliminate, those forces that do not will themselves as victorious, that are content to react. In this sense, the eternal return duplicates the very passivity of natural selection, which does not induce variations but only evaluates them. More positively, the eternal return is actively affirmed as a necessity, a necessity to be embraced and loved, to be made a self-chosen principle of life, even with all its horrifying implications.

There is something, however, of eternity that resists the will, that is the will's "most secret melancholy": the will can will only forward, into the future, it can will only in one direction, for it must take from what is given of the past. The will cannot create the past but only the future. Time itself is what limits the will, and it is only the will of the highest that can live with the horror of the repetition of the low. The will cannot conquer time, but must submit itself to time's own necessity: forward drift. The will is subject to something outside of itself, something greater, not to a God but to impersonality, the hugeness of eternity, the weight of all of the past and the open expanse of all of the future, that is, to its own limits, its own mortality. The will cannot will itself outside this constraint, this limit: "To redeem those who lived in the past and to recreate all 'it was' into a 'thus I willed it'—that alone I should call redemption. Will—that is the name of the liberator and joy-bringer; thus I taught you, my friends. But now learn this too: the will itself is still a prisoner. Willing liberates; but what is it that puts even the liberator himself in fetters? 'It was'—that is the name of will's gnashing of teeth and most secret melancholy. Powerless against what has been done, he is an angry spectator of all that is past. The will cannot will backwards; and that he cannot break time and time's covetousness, that is the will's loneliest melancholy" (*Z* 2:139).

This also makes clear how Nietzsche's understanding of time has been misrepresented as cyclical, as seasonal, as a closed circle that repeats itself when its destination returns it to its starting point. If time were cyclical, if one could wait outside its cycle, it would return to the point where it began, much like an endless tape loop. Then the question of willing the future would eventually become that of willing the past. Instead, there is a limit, an uncontrollable element of passivity: " 'That which was' is the name of the stone he cannot move" (*Z* 2:139) at the very center of willing, the limit that time imposes, which cannot be overcome but must be accepted and embraced. Time remains unidirectional, always forward, taking with it the past as it makes the future, and this will be so to eternity. It is not time itself that loops around, but the forms and configurations of matter that transform themselves, that are capable of repetition, and inevitably must, in the long run, repeat themselves as time's relentlessness pushes forward, flows into the future.

This limit to willing, the inability to will backward, this thwarting of the will's own infinity is, paradoxically, that which initiates the only genuine revenge, revenge beyond the spirit of ressentiment, a higher-order or "noble" revenge: "This, indeed this alone, is what *revenge* is: the will's ill will against time and its 'it was' " (*Z* 2:140). True revenge, unlike the petty insults perceived by the herd, can only be a revenge on time. But how can time be revenged, overcome? Only through a certain submission to its necessity, a certain reactive position that converts that necessity itself into will. The past is a series of events we do not make but inherit, or inherit even if we have made, which we must nonetheless affirm as our own in the sense that past events make us, and our overcoming, possible. To affirm a future, to affirm repetition of what will wills into the future, it is necessary equally to affirm all the accidents, events, humiliations that make willing itself the highest force. The spirit of revenge is not so much internalized as acted out: the weighty stone of the past makes carrying it into the future all the more victorious:

> All "it was" is a fragment, a riddle, a dreadful accident—until the creative will says to it, "But thus I willed it." Until the creative will says to it, "But thus I will it; thus shall I will it."
>
> But has the will yet spoken thus? And when will that happen? Has the will yet become his own redeemer and joy-bringer? Has he unlearned the spirit of revenge and all gnashing of teeth? Who has taught him reconciliation with time and something higher than reconciliation; but how shall this be brought about? Who could teach him also to will backwards? (*Z* 2:141)

Nietzsche is concerned with the "transvaluation of all values," with the genealogy and valuation of the very question of value itself. In seeking an image of the higher man, the one who has evolved beyond us, the eternal return is that which poses the greatest challenge and promise. For to live beyond the human, the all-too-human, involves the most powerful affirmation, the affirmation of the primacy of one's own becoming over one's being, of one's unknowable future over one's known or knowable past and present, even given that one's past and present are carried with their own horrors, accidents, riddles, and infirmities into that future. The indigestible (for the human) formula for this overcoming is *amor fati*, the love of fate, to affirm as necessary, and choose again, everything one wills and does: "My formula for greatness in a human being is *amor fati*: that one wants nothing to be different, not forward, not backward, not in all eternity. Not merely bear what is necessary, still less conceal it—all idealism is mendaciousness in the face of what is necessary—but *love* it" (*EH*, "Why I Am So Clever," #10).

This love of fate, this will to do again and forever what has already been done, is both to affirm the irreversibility of time (one can never undo what has been done, and thus it is unclear what a lament for the past can achieve beyond a misery in the present), that is, the unchangeability of the past, and also to recognize that this weighty past is the precise past that makes one what one is and what one can become before one can conceptualize or know it. Each "moment" of life must be affirmed as a moment one will live again and again to eternity. Yet each moment recognizes its debt to all the prior moments that condition it, and each moment affirms the open-ended effects of any present actions. This is what it is to will: to will again both the antecedents and the consequences of will itself. It is to impose the conditions of being on the movement of becoming. This is a capacity only of the strongest wills. The overman is the one who can say: I would will it all again, in every last detail. It is only this affirmation to infinity, that is, the affirmation of life lived in every detail in spite of the force of time itself, that produces the overman, that produces what value there may be today in art, in morality, in philosophy, in science.

Nietzsche contorts, twists the ties between time and matter, life and will so that these two different orders of force—one implacable, relentless, impersonal, the other mutable, particular, concrete—plunge into an interaction that subordinates the one, matter, to the requirements of the other, time. He realigns their relation to assert the superiority of the one over the other, the preeminence of the force of time over the rules of combination of matter, including those that generate life. As David Wood (2001) has recognized, Nietzsche has proposed a profound rupturing of time as presence, of time grounded in the privilege of the now by articulating the effects of time as repeatable, as a form of seriality.[14] But Nietzsche's contribution is stronger than this: not only does he fracture everyday conceptions of time and the primacy of a lived present, but he affirms above all the independent force of time, time's precedence over matter, time's force in shaping matter, not only inorganic matter, but the very matter that makes life, and thus the human, possible and capable of overcoming, becoming. Although Nietzsche has an open hostility to what he perceives is the valuation of the average, of the mediocre and the herd in Darwinism, nevertheless his work is marked by Darwin's own conception of time's positivity. It is post-Darwinian in the sense that it is precisely the Darwinian that must be overcome: the processes that generate the descent of man are also the processes that dictate the end of man and the evolution of the overman. He conceptualizes the future in terms quite consonant with Darwin's understanding of the forward force of time directing life to greater difference, greater variation, greater exploration and experimentation.

PART III. BERGSON AND BECOMING

7. BERGSONIAN DIFFERENCE

The notion of difference must throw a certain light on Bergson's philosophy, but inversely, Bergsonism must bring the greatest contribution to a philosophy of difference.—Gilles Deleuze, "Bergson's Conception of Difference"

Henri Bergson (1859–1941) is probably the last of the great metaphysicians, writing on the cusp between high modernism and positivism, conceptually located somewhere between the flourishing of phenomenology and the increasing hostility to ontology and metaphysics that developed with ever greater force through the emergence of analytic philosophy and the philosophy of science. Yet he cannot be identified with either tradition. His philosophy of life is perhaps the last unselfconscious, nonironic affirmation of the place of metaphysics alongside and as a corrective to the operations of the natural sciences. Metaphysics remains, for him, our only mode of access to that which the sciences, and intelligence itself, cannot address: the continuity of duration, the indivisible interpenetration of life and matter, the intervals between things, states, and properties—in short, the analogue continuity that marks material reality. Metaphysics is how we access the real continuity that constitutes life in its lived concreteness, and is also the means by which the living subject can locate himself or herself in and as part of the material universe as a whole, undivided by our actions. Metaphysics is that which yields us the rigorous and precise method of intuition, a particular attunement to the specificities of the real, whose details fit its object alone, whose concepts and insights are cut according to the articulations of the real itself, a method, in other words, attuned to concrete difference.[1] Bergsonian metaphysics, immensely popular between the publication of *Matter and Memory* in 1896 (1988 [hereafter MM]) and *Duration and Simultaneity* in 1922 (1965), was subjected to bitter and often unfair criticism from a wide variety of detractors, and to the extent that it was a powerful and popular movement in Bergson's earlier career, it was reviled and ridiculed more or less from this time on, functioning largely as a historical anachronism, a curiosity, more than as a dynamic force in contemporary philosophy.[2] Bergson's work today, though, is not without its contemporary readers, who with some effort have managed to demonstrate his ongoing relevance to the natural sciences as well as to philosophy, especially philosophy interested in the question of the virtual and the future.[3]

Bergson's writings demonstrate no evidence of having read Nietzsche, as Nietzsche himself never read Darwin; nevertheless, his understanding of duration and creative evolution brings together the key insights of Darwin, modulated by a Nietzschean understanding of the internal force of the will to power and the external force of the eternal return. The will to power is transformed in Bergson, not into a will to command or obey, but a will, a force, or *élan vital*, which propels life forward in its self-proliferation. Bergson must be regarded not only as the most philosophically rigorous of the early twentieth-century Darwinians, but primarily as the philosopher most oriented to the primacy of time, time as becoming, as open duration. His concern with the relations between life and matter is closely entwined with the interests of Darwin. Yet, unlike Darwin, whose aims are to fill in the details, through scientific observation, of a picture of biological emergence and development, Bergson is concerned with finding the appropriate limits of scientific analysis, and with what it is that scientific methodologies must leave out. His project, like Nietzsche's, is fundamentally philosophical. It is not concerned with empirical details (though it does not shun the use of such research); rather, it is a conceptual reflection on the accomplishments and limits of the sciences, concerned with the production of its own unique concepts, which are not reducible to the aims and principles of the sciences but perhaps are required by the sciences if they are to gain self-understanding. He does not develop either a philosophy of science, that is, the philosophical analysis of scientific texts and experiments, or an epistemology, that is, an account of how knowing is possible and how scientific forms of knowing know their objects. Bergson's concerns are different. Instead of making philosophy a form of passive acceptance of the givenness of the discourses and practices of the sciences, he makes it a productive concern that both functions *alongside* the sciences, operating with different aims and methods but able to make collateral use of the sciences as much as of the arts, and also functioning as it were *underneath* the sciences, making explicit their unacknowledged commitment to philosophical and ultimately ontological concepts.

Bergson's relations to Nietzsche are perhaps more indirect and difficult to understand; nevertheless, he shares some of Nietzsche's reservations about the problems of evolutionary theory, particularly in the hands of the social Darwinists. There is not a direct, historical connection between Bergson and Nietzsche. But there are certain points of resonance or similarity between their work, as well as elements in tension or uneasiness which must also be acknowledged. In particular, like Nietzsche, Bergson wishes to elaborate a theory of time in which the past is not the overriding factor, and in which the tendencies of becoming that mark the present also characterize the future. As I argued in the previous chapter, Nietzsche does not have a circular conception of time: the eternal return is not the return of a seasonal, cyclical rhythmicality (with which it is commonly confused), for it is an imperative for the future, a future that is in continuity, through divergence and elaboration, that is, through difference from rather than through any linearity, causal or otherwise, with the present. The eternal return is the imperative that dictates the structure of affirmation, the affirmation of the weight, the enormity, of a commitment in the future, a promise, which is both the forgetting of the past in order that a future be developed beyond it and the reconciliation with the past in which the promise is made and from which the resources for keeping it are formed. In this sense, Nietzsche is Bergsonian perhaps more than Bergson is Nietzschean: insofar as the future functions as a mode of unpredictable continuity with the past, the future springs from a past not through inevitability but through elaboration and invention. If Nietzsche is in this sense Bergsonian, though, it is significant that Bergson is *not* Nietzschean: his metaphysics does not contain in itself an ethical and evaluative project but returns to the ontological roots of the Darwinian schema. Or rather, it is an ethical project but not an interrogation of the value of value.[4] In this sense, his interests in Darwin and Darwin's implications for understanding philosophy are probably closer to the cosmological Nietzsche than they are to Nietzsche's morality.

Bergson is interested above all in elaborating the distinction between, and the intermingling of, mind and matter, and how they implicate the operations of time and space respectively. Like Darwin, Bergson is interested in the processes of development, processes that induce change, that demonstrate time's forward direction, and in a future that is based on the resources of the past while it inevitably surpasses them, that involve innovation, emergence, and the creation of the new and the unforeseen. Yet, like Nietzsche, he is concerned to see life itself as an active dynamism, a series of forces, that function with their own pragmatic links to time; he affirms life, and its possibilities of becoming, as a supreme value. Neither directly Darwinian nor Nietzschean, nevertheless Bergson must be understood as a kind of natural child of their quite mixed heritage, a philosopher concerned with the place of life amid matter, with the future of life, the force of life, and with its conceptual and epistemological limits.

As was the case in my readings of Darwin and Nietzsche, it is quite clear that Bergson's vast and underappreciated writings cannot be adequately addressed here. I can focus on only three key elements of Bergson's contributions to ontology: first, his understanding of matter and its relation to memory; second, his account of the relations between past, present, and future as they manifest themselves in evolutionary movement; and third, his understanding of the distinction between the virtual and the actual and the relation between the virtual and the practice of intuition. With these three components, we have enough to extract elements that may be constructive in retheorizing time as force, and in devising political strategies that make the most use and value of the strange and non-self-evident character of time. With this aim in mind, though I refer to his other writings where they are relevant, I concentrate primarily on the two key texts focused on duration, *Matter and Memory* and *Creative Evolution*, his most famous and misunderstood writings, devoting this chapter to the earlier text and the next to *Creative Evolution*.

Differences in Kind and Differences of Degree

If, as Deleuze claims in this chapter's opening epigram, Bergson provides a crucial, indeed, the "greatest contribution" to the philosophy of difference, it is because his writings link metaphysics to the discernment of differences in nature, natural difference or differences in kind. Bergson seeks to distinguish two kinds of difference (differences that are themselves differences in kind): between differences of degree, which he construes as differences of magnitude, quantitative differences, differences of more or less, measurable differences; and differences of nature or in kind, which are qualitative differences, differences impossible to measure or describe in numerical terms but discernible in and for conscious mental life and experienced in the continuity and ever-changing movement of duration. If quantitative differences are measurable, they indicate spatial differences, differences external to each other, differences between things, differences that can be marked or characterized through their coordination with a third term, a number, that is, through measurement.[5] Such differences are discrete, discontinuous, and homogeneous. They can be divided in infinite ways without transforming their nature.

Differences in kind, by contrast, can be construed as internal or constitutive difference, continuous, heterogeneous, interpenetrated, without clear-cut outlines or boundaries, and incomparable with each other or with a common measure. We can divide these differences only into successive interpenetrating wholes rather than into juxtaposed parts. These differences cannot be measured, or if they are, they are reduced to external differences through the mediation of a third or measuring term. Their only real measure is a metric that is unique to each division, each particularity, each "moment." The interaction of "parts" (if they can be understood in these terms) produces a flux that is not without some order or organization, but whose structure fluctuates and transforms itself over time. At any natural division, it forms an indivisible totality. Although these differences and divisions cannot be measured by some outside term, they can be intuited, discerned with careful attention to the natural articulations of the real, even if they cannot be designated or represented without reduction to externality or to quantity. Qualitative differences are internal differences, differences constitutive of the particularity of events: differences in kind always immerse themselves and are invested in the movement of duration itself, which is that very movement of differing from itself, the movement that ensures that nothing retains absolute self-identity over time, even if it may, through artificial cuts, if time is frozen or rendered synchronic, retain a measure of self-resemblance and cohesion, that is, a measure of spatial integrity.

Among the tasks of philosophy is to adequately distinguish between these nuanced differences or multiplicities, to ascertain how one difference (of degree) covers over and hides another (in kind), and to show what is left out, what is unrepresented or uncharacterized about differences in kind, differences of nature, in our dominant scientific and philosophical frameworks which focus on quantitative differences. Quantitative differences can, in principle, be compared with each other insofar as they elicit a common, abstract, or infinite measure that distinguishes them in that single respect. Qualitative differences, by contrast, are potentially incommensurable: they need to be what they are (the same) in order to provide a stable basis of comparison, for comparison is itself the spatialization, the placing side by side, the rendering contiguous of two or more things, processes, or qualities. Qualitative differences are thus incomparable, unique, lacking self-identity, for they differ not only from quantitative differences and from any stable system of measurement but also from themselves. Qualitative differences are internal differences which ensure that, if duration is real, no term can remain what it is but differs from itself as time progresses.

It is significant that this difference between different conceptions of difference underlies not only an ontological rift between the quantitatively oriented differences of degree and the qualitatively directed differences in kind, but also a political difference between strategies oriented toward the affirmation of suppressed identities and those directed to the affirmation of incomparable differences. The first strategy utilizes quantitative difference, differences of degrees, differences in which one category of subject (women, homosexuals, ethnic or political minorities) is relegated to the status of lesser or greater relative to another category (men, heterosexuals, ethnic or political majorities). The second strategy utilizes the so-called politics of sexual difference associated with Irigaray, and with Derridean deconstruction, which are both invested in noncalculable differences, differences-to-come rather than merely with distinctions and oppositions that presently exist. Sexual difference is not a measurable difference between two given, discernible, different things—men and women, for example—but an incalculable and continuous process, not something produced but something in the process of production. The first sense of difference is oppositional, binary, dichotomized: difference is defined through negation, absence, and lack. Women are defined as nonmen, lacking the characteristics that make men men. Oppositional differences produce discontinuities, gaps, boundaries, bounded entities divided by a logical barrier or gulf. What constitutes categories is the presence or absence of valued or denigrated characteristics or qualities: to have or to not have particular or given attributes. These differences resolve themselves into diverse forms, but diversity is always construed as comparative, as degrees of attainment of given attributes. Capable of being placed alongside each other, these differences can be measured, monitored, surveyed, assessed, even redistributed. But the value of the given characteristic by means of which they are compared remains unquestioned. And the processes by which these characteristics vary and differentiate themselves remain unelucidated.

In the second sense, difference is neither oppositional nor complementary; differences here are understood as occupying different conceptual landscapes, qualitatively different, and thus incapable of being specified in advance or compared to each other, for these are differences that are in the process of being made rather than already given. Sexual difference and racial difference cannot be understood productively *except* in terms of such internal difference, for they cannot be understood as the comparison of two or more already known and measured sexes, two or more given races, categories or groups. Rather, they can be represented only as *yet-to-come*: what woman might be, what man can become, what races are in the process of becoming, which cannot be known in advance or definitively and is incapable of being measured.[6]

These two kinds of differences in Bergson, the difference between that which remains the same, which does not differ from itself (i.e., matter), and that which does differ from itself (life, duration) can be more directly and straightforwardly elaborated as the distinction or opposition between objects or things locatable in space, which are capable of measurement, regulation, and repetition; and sensations and affects, which always vary or transform themselves over time, through duration and its movements of continual elaboration.[7] Sensation, consciousness, mind, or life—all in some ways interchangeable terms in Bergson's oeuvre—transforms itself in quality rather than magnitude; it is that which never remains the same as itself, that which cannot be cut out of its continuum to be analyzed, that which varies from itself.

Sensation cannot admit of degrees; sensations are not more or less intense: we do not experience sadness as a muted mourning or happiness as an intensified joy, even if we sometimes speak in these terms. Sensations cannot be understood as intensive magnitudes. Magnitude, degree, number are usually understood through a spatialized relation of containment: greater magnitudes contain smaller ones, and smaller ones can be added together to create larger magnitudes. But how can sensations, which are never the same as themselves, be understood as larger or smaller, greater or lesser than each other? How can one sensation (the greater magnitude) be understood to contain another (the lesser magnitude)? Bergson devotes a good deal of attention to discussing how magnitudes have come to be attributed to sensations through the collapsing of the external cause of a sensation (which, located in a specific object or relation, can often be measured) with the effect, the experience of the sensation. Our experience of luminosity, for example, can often be correlated with quite precise measurements of luminous sources, but the experience of light or color is itself always qualitative: a lighter gray is not the same gray muted but an altogether different, unique, sensation.[8] Sensation changes in nature but not in magnitude, and these changes cannot be charted or mapped spatially, for such spatialization places terms outside of each other (and thus capable of measurement, of reduction to the quantitative). These changes in sensation interpenetrate each other and differentiate themselves only in duration. Sensations are always in continuous variation: the continuity of a sensation does not simply extend it in time but transforms its quality. Numbers, the measure of magnitude, are always, by contrast, constant and discontinuous (1959, *Time and Free Will* [hereafter TFW], 82–83; see also MM 41–43), whereas qualities are never discontinuous for they blur into each other imperceptibly:[9]

> Everything is not counted in the same way, and . . . there are two very different kinds of multiplicity. When we speak of material objects, we refer to the possibility of seeing and touching them; we localize them in space. In that case, no effort of inventive faculty of symbolical representation is necessary in order to count them; we have only to think them, at first separately, and then simultaneously, within the very medium in which they come under our observation. The case is no longer the same when we consider purely affective psychic states, or even mental images. Here, the terms being no longer given in space, it seems *a priori*, that we can hardly count them except by some process of symbolic representation. (TFW 85–86)

Quantitative and qualitative differences, the differences between relative magnitudes and irreducible qualities, indicate two broad directions or ontological tendencies in Bergson's understanding: correlated with and supporting quantitative or magnitudinal differences is the world of matter, understood as inert or reactive; framing and contextualizing all qualitative differences is the world of memory, sensation, consciousness or life, systems of relative freedom or indetermination that introduce surprise and unpredictability, duration, into a world usually understood in terms of causal predictability as a pure or continually remade present. If matter is capable of quantitative considerations, this is because matter finds its natural milieu in extension, in spatialization, in the possibility of having its parts spread out externally, side by side or synchronically, its nature being revealed at any single moment in time. Memory, sensation, consciousness—qualities only of living beings (and by no means the privileged object of man alone)—involve the past's persistence in the present, the power of transformation that ensures that objects, and especially subjects, are not what they once were, but are in the process of becoming more. In Bergson's early work, matter is naturally the object, or rather, the content, of quantifiable, extended, mappable space, whereas memory is the content or object of qualitative duration or becoming.[10]

Matter and memory, the present and the past, space and duration, the inorganic world analyzed by physics, and the psychical world of lived experience are all different names for or angles on this fundamental opposition between quantitative and qualitative multiplicities, differences of degree and differences in kind. Although Bergson is commonly understood as an irredeemable dualist, for whom binary oppositions, such as mind and matter, are given, his position is more complex and less easy to decipher than oppositional models allow. The distinction between these two kinds of multiplicity or difference is itself qualitative and nonoppositional; the terms provide extremes of a continuum, a difference of tendency or impetus, an "endosmosis," that is, a difference of degree. Matter will turn out (in Bergson's more mature work) to be memory in its most dilated form; memory will be understood as the most contracted expression of matter; space in its global or cosmological form becomes, ages, has a history, is subjected to duration; and time itself is the condition of the simultaneities that contract to constitute space.[11] The difference between differences of degree and differences in kind itself becomes a difference of degree.

To understand how these terms undo each other, entwine with each other, constitute the blended mixture that makes up all experience and materiality in its totality, we need to look in more detail at the relations between matter and memory, the encounters that constitute the field of evolutionary becoming.

Perception

The strange inventiveness of Bergson's conception of the relations between matter and memory, or, in more conventional philosophical terms, between body and mind, becomes apparent at the very opening of *Matter and Memory.* He defines matter, not in terms of substance or extension, as it has been generally understood in the Cartesian tradition, but in terms of images: matter is the ongoing production or profusion of images. The structure of matter is imagistic, which is not to claim that it is reduced to the imagistic perception of a subject (i.e., idealism) or that the image is necessarily or in any privileged manner visual. Matter is conceptualized as midway between the image, so central to idealism, and the object-in-itself, so central to materialism. Matter is an aggregate of images that occasion, in the presence of a perceiver, a series of multisensory perceptions, images capable of representation by many if not all of the senses and by other perceivers: "Matter, in our view, is an aggregate of 'images.' And by 'image' we mean a certain existence which is more than that which the idealist calls a *representation*, but less than that which the realist calls a *thing*—an existence placed half-way between the 'thing' and the 'representation' . . . the object exists in itself, and, on the other hand, the object is, in itself, pictorial, as we perceive it: the image it is, but a self-existing image" (*MM* 9–10).

Matter is a multiplicity or aggregate of images. This is both a form of realism (insofar as the object exists in itself, independent of any observer or subject) and a mode of idealism (insofar as matter coincides with and resembles its various images). Yet Bergson's position cannot be identified with either realism or idealism, for he claims that both share a mistaken belief that "perception has a wholly speculative interest: it is a pure knowledge" (*MM* 28). To perceive, for both idealism and realism, means to know, to receive a disinterested registration of a pure knowledge—either (for realism) impressions that require augmentation through greater precision or (for idealism) impressions that provide an absolute connection to objects. But for Bergson, perception cannot be equated with knowledge, for it is primarily concerned with action. Perception is the way living beings deal with matter, utilizing the images that are the world itself for their needs and activities.

His doctrine of the imagistic nature of matter is linked to this fundamentally pragmatic understanding of our relations to matter.[12] The universe is an aggregate of images. These images act and react with each other according to relatively predictable principles, which we describe through scientific or natural laws. The images that constitute the material universe, in their law-like operations, are in principle perfectly predictable, which means, following a Laplacean model, that a super or divine intelligence, a god or demon, which could somehow picture all of the interactions of these images, would be able to predict their future. Their future is already contained in their present, just as the present could have been predicted from a perfect knowledge of the past.

Among these images that constitute the materiality of the universe is the image of my body, a material object like all others, except in one respect. Whereas the images that constitute the universe can be known only from outside, through perception, the image(s) that constitute my body are capable of being known from within, through affection. My body occupies a privileged position insofar as it is a moving center through which I gain access to and perception of all the other objects and is thus a continually reorienting framework through which objects are contained or represented in a field surrounding it, a context.[13] The body mediates between the impact of external images and the transmission of movement back to those images; unlike inorganic objects, living bodies act as a kind of storehouse for energy, containing within themselves, in varying degrees, the possibility of choosing when and how to act and react: "My body, then, is the aggregate of the material world, an image which acts like other images, receiving and giving back movement, with, perhaps, this difference only, that my body appears to choose, within certain limits, the manner in which it shall restore what it receives . . . *My body, an object destined to move other objects, is, then, a center of action*" (MM 19–20; emphasis in original).

This capacity to store up energy and to discharge it in order to maximize the utility of one's actions is perhaps one of the most elementary functions of life itself, and a measure of the degree of freedom that all life exhibits. It is the capacity to initiate something new and unpredictable, something not contained in antecedents. The living body is capable of acting and reacting on other images, effecting transformations in their functioning in any number of ways, depending on its interests, needs, desires. In short, through its capacity for choice, for a decision among a number of possibilities that matter offers it, the living body introduces surprise into the universe, produces arrangements and interchanges with matter that have never occurred before nor perhaps will occur again, that were in no way already contained in matter.[14] What the body and its capacity for the perception of objects, including other bodies, engenders is the orientation and transformation, the framing, of the material universe so that it helps to facilitate the actions initiated by the living being. Perception, then, must be linked to nascent or dawning action, action-in-potential. Perception is not a passive knowledge, the reception of the impress of the material images, that is, sense data; rather, it is the filtering and sifting through the myriad properties of objects to find those qualities that interest life. If perception impels us toward action and thus toward objects, then to that extent objects reflect my body's possible actions upon them. My body serves to filter, simplify, highlight, or outline those qualities of the object that may be of relevance or use. This does not occur in the form of a conscious or unconscious judgment but is inherent in the very act of perception itself, which is always a simplification of the object, the elimination from it of what does not interest us.

The difference between matter and perception is thus a difference in organization, not in kind: if matter is nothing but an aggregate of self-subsisting images, then the perception of matter is not a higher-order image—the image of an image, the subjective impression of an object's properties—it is the same images oriented toward the organizing force of a central image, the image of my body. The difference between matter and perception is not simply the difference between an object and a subject but a difference in the potentiality or mobility of images. The subject is a peculiar sort of object, linked through the body's central organizing position, to frame and make use of the rest of matter. "Like a compass" (*MM* 23), my body is a moving, dynamic object among all the others that make up the world, which continually changes the position of objects according to the relativity of my movements. What differentiates my body from other objects is, in the first instance, the way the image that is my body has a peculiarly privileged relation to action. My body is distinguished from other objects not because it is the privileged location of my consciousness but because it performs major changes in other objects relative to itself, because it is the central organizing site through which other images/objects are ordered.[15] The image that is my body occupies the center of the material universe, and as it moves and changes, it brings about a kaleidoscopic shift in the orientation of that universe. If the universe in itself and outside the body's perception is subjected to perfectly predictable action, how is it that the living body can render this measurable, causally connected universe unpredictable?[16] How do these two types of image, one with a center (my body) and the other without any center (the universe), one animate and the other inanimate, coexist? How is it that subjects are inserted into the determinate order of objects? How can these living subjects transform objects and introduce into them the activities of indetermination? In other words, what is the relation between mind and matter, and what is the manner of their coexistence?

Bergson sees this relation as one of mutual occupation. Scattered throughout the system of linked images that constitute the material world are living systems, centers of action, zones of indetermination, points where images are capable of mobilizing action by subordinating other images to the variations and fluctuations, changes of position and perspective afforded by these centers of action. Life can be defined, through a difference in kind from matter, by the necessity of prolonging a stimulus through a reaction, through the voluntary capacity to store energy instead of immediately expending it. The more simple the form of life, the more automatic the relation between stimulus and response. In the case of the simplest of living organisms, the protozoan, the organs of perception and the organs of movement are one and the same. Reaction seems like a mechanical movement, an automatic response. Even here, however, the protozoa exercises a measure of freedom or indetermination: it exercises a small degree of freedom in its contractile possibilities, in the "choices" it exercises over when to contract or to expand, what relation it has to external objects. In the case of more complex forms of life, there is interposed both a delay, an uncertainty, between a perceptual reaction and a motor response, and an ever-widening circle of perceptual objects which in potential promise or threaten the organism, which are of "interest" to the organism and which it can connect to the object perceived. Bergson's claim seems to be that the more complex the form of life, the more unpredictable the response, the more interposing the delay or gap, the more freedom, and the greater consciousness.

This notion of life, mind, perception as both the organization of images around a central nucleus and as the interposition of a temporal delay between stimulus and response distinguishes Bergson's position from any form of humanism and from charges of anthropomorphic projection. Mind or life is not a special substance, different in nature to matter. Rather, mind or life partakes of and lives in and as matter. Matter is organized differently in its inorganic and organic forms; this organization is dependent on the degree of indeterminacy, the degree of freedom, that life exhibits relative to the inertia of matter, the capacity that all forms of life, in varying degrees, have to introduce something new. This something new, a new action, a new use of matter, a new arrangement or organization, is brought into existence not through complete immersion in matter but through the creation of a distance that enables matter to be obscured, to be cast in a new light, or rather, to have many of its features cast into shadow.

It may be for this reason that Bergson develops one of his most striking hypotheses, especially in light of the contemporary fascination philosophy has with cognitive science and neurological models: the brain does not make humans more intelligent than animals, the brain is not the repository of ideas, of mind, of freedom or creativity. It stores nothing, it produces nothing, it organizes nothing. Yet, it is still that which partially explains or conditions the possibility of innovation, creativity, and freedom insofar as it is the means by which a delay is interposed between stimulus and response, perception and action, a capacity for rerouting and reorganizing the perceptual-motor circuit:

> In our opinion . . . the brain is no more than a kind of central telephone exchange: its office is to allow communication or to delay it. It adds nothing to what it receives . . . That is to say that the nervous system is in no

> sense an apparatus which may serve to fabricate, or even to prepare, representations. Its function is to receive stimulation, to provide motor apparatus, and to present the largest possible number of these apparatuses to a given stimulus. The more it develops, the more numerous and the more distant are the points of space which it brings into relation with ever more complex motor mechanisms. In this way the scope which it allows to our action enlarges: its growing perfection consists in nothing else. (*MM* 31)

The brain intercedes to reroute perceptual inputs and motor outputs. It links, or disconnects, movements of one kind (sensory or perceptual) with movements of another (motor). The brain functions, in Bergson's conception, not to produce images or to reflect on them, but to put images directed from elsewhere, from the world, into the context of bodily action. The more developed the organism, the wider in nature are the perceptual or sensory inputs and the broader the range of objects that make up the scope of the organism's action. The brain enables a gap or delay between stimulus and response which in turn enables, but does not entail, a direct connection between perception and action. The brain enables multiple, indeterminable connections between what the organism receives (through perception or affection) and what is available for it to act on, making possible a genuine freedom from predictability. Freedom in his conception is neither the absence of causes (as traditional proponents of free will assert) nor the range of options or possibilities available (free choice) but the capacity to act in concert with one's past to bring about a future not contained in it.[17] It is precisely this delay or interval that lifts the organism from the immediacy of its interaction with objects to establish a distance that allows perceptual images to be assessed and function in terms of their interest, that is, their utility or expedience for the subject. This interval serves as a kind of principle of selection of those elements of the object that link it to the interests of the living being.

The object thus can be understood to contain both real action, the indiscriminate action of its various features on whatever surrounds it and comes into causal connection with it, and virtual action, that potential to exert specific effects on or by a living being of the kind that the being seeks or that may interest it. This cerebral delay allows the object's indiscriminate actions on the world to be placed in suspension and for the living being to see only the relevant or harnessable properties of the object: "To obtain this conversion from the virtual to the actual, it would be necessary, not to throw light on the object, but on the contrary, to obscure some of its aspects, to diminish it by the greater part of itself, so that the remainder, instead of being encased in its surroundings as a *thing*, should detach itself from them as a *picture* . . . There is nothing positive here, nothing added to the image, nothing new. The objects merely abandon something of their real action in order to manifest their virtual influence of the living being upon them" (MM 36–37).

The zones of indetermination introduced into the universe by life produce a kind of sieve or filter on the images of the world, diminishing the full extent of the object's real effects in the world in order to let through its virtual effects. What fills up this cerebral interval and interposes itself between sensation and action to enrich and complicate both are affections, body-memories (or habit-memory), and pure recollections (duration), the qualitative elements constituting life. By their interposition, they become "enlivened" and capable of being linked to nascent actions, drawn out of their inertia. Through them, objects are put into new contexts, utilized in new ways, produce new effects. Inventiveness is introduced into the rigid determinacy of matter's relations to itself.

Memory

Bergson speaks of two different kinds of memory. One is bound up with bodily habits, and thus essentially forward-looking insofar as it aims at and resides in the production of an action, however habitual. This habit-memory is about the attainment of habitual goals or aims: driving a car, typing, activities in which the body "remembers" what it is to do without conscious intervention, yet that once needed to be consciously learned before being automatized. It has a kind of "natural" place in the cerebral interval between perception and action, for it is the most action-oriented, the most present- and future-seeking of memories from the inert past. It consists in habits, previously acquired, and automatized chains of action that filter the real details of objects in order to highlight what in them is, or has been, of direct utility. Acquired by repetition, synthesis, and schematization, habits contract a series of closely linked and regulated activities into an initial impulse, which then sets off the automatic or habitual chain of behavior. Habit-memory synthesizes a series of repetitions into a given form that is relevant to behavior in the present. As Bergson claims, habit-memory is a past that "is lived and acted, rather than represented" (MM 81), a series of mechanisms stored from the past, waiting for activation in the present: "In truth it no longer *represents* our past to us, it *acts* it; and if it still deserves the name of memory, it is not because it conserves bygone images, but because it prolongs their useful effects into the present moment" (82).

If habit-memory repeats the past in the present, memory proper recalls it, represents it, just as perception represents the material image. For Bergson, this distinctive recollection of the past occurs only when our attention is drawn away from the present and immediate future, when our attention is in a state of relaxation, or makes a specific effort to direct itself to the past. The past itself is "fugitive" (*MM* 83), fleeting, accessible only through the movement of turning away from the present. Memory proper, recollection or remembrance, must be understood as always spontaneous, tied to a highly particular place, date, and situation, unrepeatable, singular, unique, perfect in itself (incapable of developing). If habit-memory is future-oriented, memory proper is always and only directed to the past. Where habit-memory interposes a body schema between sensation and action, memory proper is directed toward an idea. If the cerebral delay could be indefinitely postponed, Bergson suggests, these memory-images, precise, concrete images from the past, would serve to fill the breach. This, of course, is precisely what occurs in the case of sleep, which severs the impetus of the perception from the requirement of action and can thus more readily tolerate the interposition of detailed and highly particular memory-images, which serve no practical function.

As he implies, perception always inclines us to the future; it is only because there is a delay or rift between perception and its future motor action that this orientation to and relevance of the past is possible. Movement and action drive the memory-image away, but equally, perception and action in the present gain their liberty, their capacity for innovation in the future through the unexpected intervention of memories, which enable this present to be cast in an unexpected light. The present is fractured or nicked only by the past. Yet habit-memory, the impulse of the past to prefigure the present in terms of what is familiar to it and already accommodated by it, with its focus solely on the present and imminent future, that is, with its orientation always to adaptation, drives away, overpowers the more languid process of reverie or reminiscence needed to summon up or actively enter the sphere of the past that is recollection.

It is almost as if the past has two alternatives by which it can exercise some influence on the present. In the first, it can contract itself into the preparation for muscular movements of the body, in which case, traces of the past reappear in the present in the form of habits, automatized motor mechanisms that schematize past behavior for a present use. In the second, it appears in the form of memory-images, which represent or picture past events in their detailed specificity. The first alternative appears through conscious effort, through repeated learning; the second occurs spontaneously and often unbeckoned, as "capricious in producing as it is faithful in preserving" (*MM* 88).

The act of recognition is the point at which memory proper and action are at their closest point. Recognition involves the correlation of a current perception (or perceptual object) with a memory that resembles it. But recognition is not simply a correlation, for recognition would be guaranteed to occur whenever there were memory-images and would be abolished whenever they were missing (an explanation that cannot take account of the phenomena of psychic blindness and the detailed and selective forgetfulness of aphasia). Recollection, and thus recognition, occur when a memory-image that resembles a current perception is carried along with the perception by being extended into action; but it is significant, Bergson claims, that even in the absence of a memory-image, there may be the possibility of recognition. It is not simply the summoning up of a prior image to correlate with or correct a present perception, for this makes present perception disinterested and the memory-image it resembles merely passive, awaiting retrospective recall. Both the perception and the memory-image have a certain investment in activity. Recognition of what is familiar enables us to outline or schematize our sensory impressions, to act with a minimal effort and consciousness on present objects.

Bergson illustrates with the example of walking around an unfamiliar neighborhood. When I walk there the first time, everything is new; I am conscious of every orientation and unfamiliar with each possible direction. My movements are discontinuous and without conviction or confidence. When I have walked my route a number of times, when I have familiarized myself with my environment, I can react automatically now to that which I hesitated over before, strolling in whichever direction with a certain confidence in my movements, without even a conscious awareness of passing certain objects or landmarks. My familiarity, my capacity to now recognize my environment has been compacted into the range of readily assumed movements available to me.

Recognition can in general be understood, not as the meeting point of the past or memory and the present or perception, but as the interposition of habit-memory, action-oriented triggers for behavior, between a pure memory, which has no interest in current action, and a perception, which is interested only in current action but is likely to be overwhelmed by the number of images an object generates without the schematizing/synthesizing power of habit.[18] This is why, for Bergson, the dog recognizes his master although it seems likely that animals access the past only through habit-memory rather than through the elaborated imagistic and representational structure of memory proper.[19] This also explains why the various disorders of memory that so fascinate Bergson[20] do not involve the loss or destruction of memory-images (a claim he believes is incoherent, for it assumes that memories are stored somewhere in the cerebral apparatus), but rather with a partial, temporary or permanent impairment of the motor mechanisms by which these memories can be linked to current actions. These are not, strictly speaking, failures of memory, but rather failures of habit, the breakdown of the bodily schema that enables the performance of a habitual task, lesions or disorders that effect the enactment of possible action.[21] Those memories positioned more closely to the habit pole of recollection function in conjunction with or as an adjunct to perceptual innervations extending out to impending actions: this is the closest memory comes to activation, to touching and influencing the present.

In this sense, recognition is not so much thought, conceptualized, a matter of knowledge, as acted, put into play, a motor adaptation. An intellectual recognition, which is to be distinguished from automatic or habitual recognition and for that reason is considerably more rare, is always an act of attention, an act of discerning in a current perception the concrete memory it calls to mind. Recognition itself runs in two forms, according to whether it is habit-memory or memory proper that is involved. The recognition that occupies us in everyday life comes from an inattention to the specificity of the images recognized; an active recognition involves an effort of attention, an active seeking out of memory-images and the establishment of a reflective connection between them:[22] "Every *attentive* perception truly involves a *reflection*, in the etymological sense of the word, that is to say, the projection, outside ourselves, of an actively created image, identical with, or similar to, the object on which it comes to mold itself" (*MM* 102). Recognition, then, is divided into the two poles of memory: generally, perceptions tend to associate more readily with habit-memory, though through an effort of attention they can connect with the more detailed, elaborated, and specific images provided by memory proper.

Insofar as memory images can insert themselves successfully, it is difficult if not impossible to distinguish the (current) perceptual component from the memory images that augment and enrich it. This is why Bergson acknowledges that although "pure perception" and "pure memory" are ideal limits or orientations, in everyday life we find only their varying mixtures, forms of coexistence, perception infiltrated by memory, memory seeking expression in current activities, current activities summoning up relevant or associated memories.

If memory is either activated or drawn out of the past by a current perception, or if memory seeks out and selects a perception that resembles it, it is carried along the path to action. Its virtual elements are melded into the activity or labor of the perception itself and a mixture, an "impure" perception, is produced. It is not clear, even with reflection, whether we can accurately separate the elements derived from the present from those from the past: they form a complete and seamless whole.

It is significant that, for Bergson, if there is a movement from memory to perception in acts of recognition, we can also pass in the reverse direction, from movements to memory, a movement that is needed to "complete" the perception of the object, which has been stripped of its manifold connections in reality to serve as a point of interest for current perception and impending future action. Between the current perception and the interposed memory there is a "mutual tension" that both enables them to intervene into the operations of the other but that also stabilizes and holds the past and the present apart. Memory returns to objects the rich potential they have for functioning outside their familiar uses; it returns to them the qualities, properties, contexts that perception must eliminate in order to act on the object. Perception can never be free of memory and is thus never completely embedded in the present, but always retains a reservoir of connections with the past as well as a close anticipation of the imminent future. The present is extended through memory into the past and through anticipation into the near future.[23]

This movement from the multiple circles of memory must occur if a productive circuit between perception and memory, where each qualifies the other, is to occur, that is, if there is to be the possibility of a reflective perception or a directed recollection. Bergson conceptualizes this circuit in terms of a return movement from the object to recollection, in increasingly concentrated or dilated circles. There is a fundamental solidarity between the object of perception and the circuits of memory that enables us to elucidate and elaborate that perception when we concentrate on or pay attention to the object. These different circuits, or planes, of memory may share nothing in common with each other besides the resemblance or association of each with the object. Bergson thinks of memory as fundamentally elastic: it is capable of existing in a more or less contracted or dilated state.

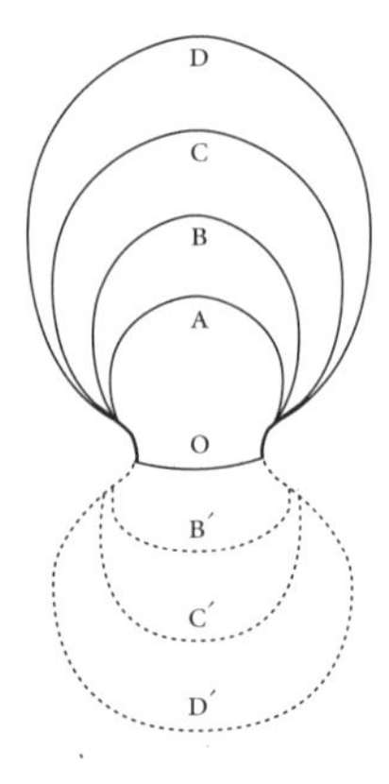

Figure 1
Source: Henri Bergson, *Matter and Memory*, trans. N. M. Paul and W. S. Palmer (New York: Zone, 1988), 105, fig. 1.

The whole of memory is contained within each circuit in concentric degrees of concentration (see figure 1).

In the circuit from the object to recollection, one does not seek out simply an image to resemble the perceptually given object; these different circuits cluster and are formed concentrically around images connected by resemblance to the object, which exist in more and more dilated form as they are removed in immediacy from the object. For example, the smallest circuit, A, consists in the object's (O) afterimage. As the circuits widen (in B, C, D), the memory becomes deeper and more detailed, more removed from action, and more filled with content and context. Memory-images become richer and conform in their detail more to the object's perceptual image. The object's perceptual images (A, B, C, D) become deeper and more complex, partly as a result of the virtual images, memories (B′, C′, D′), now carried along with and in the object.[24] To move from one circuit to the next cannot be accomplished directly, for each time one must return to the present to be able to leap once again into the medium of the past. If, in the case of mechanical recollection, the past, in the form of habit-memory, moves toward action, in the case of attention, the past is not just directed toward action but is expanded and grows richer, confirming the perceptual image.[25]

Each of the concentric circuits is a different degree of contraction or dilation of the past. They represent not only different degrees of detail and elaboration—those closer to the object are more tied by tension, through concentration, to the outlines of the object, functioning indistinguishably from the object itself, while those further away, drawn through a more elastic elaboration, are more immersed in the detailed recollections that make up the more personalized past of the perceiver and are thus more removed from the object's outlines. They also draw into the circuits of the object a variety of associations, memories, connections that free the object more and more from the constraints of its unattended or unnoticed, and thus highly schematized, existence.

Memory thus donates to the object the potentiality or virtuality of the past, which helps restore to it or replaces for it what perception necessarily must strip away. In exchange, the memory-images that function to highlight or illuminate the object from unexpected angles abandon part of their virtuality in order to actualize themselves through their attraction to present perception and its accompanying movements. Bergson describes the movement from the object to memory as centripetal, and from memory to object as centrifugal. These two opposed forces, the real actions of the object and the virtual actions of the memory-image, converge to bathe the object in its potential, to make it available for present and future use in ways unrecognized by habit.

Past and Present

If perception is bound up with and dominated by the present and the impending future, and memory is linked to the past, unless it can somehow associate itself with a current perception, we have fundamentally a difference in kind between perception and memory and between the present and the past. Bergson's understanding of duration as continuous, dynamic, and interpenetrating change resides primarily in his conceptualization of the complex and misunderstood relations between past and present. The relations between past and present are clearly of central concern not only to all the humanities, for whom the past can never be understood simply as an inert backdrop for contemporary concerns, but whose influence persists, and transforms itself over the passage of time; they are also of relevance to the natural sciences, which attempt as much as possible to control, regulate, and reduce the movement of time.

The present is that which acts and lives, that which functions to anticipate an immediate future in action. The present is a form of impending action, a way of acting with a view to what is next. The past is that which no longer acts, and although it lives a shadowy and fleeting existence, it still *is*, it is real. The past remains accessible in the form of recollections, either as motor mechanisms in the form of habit-memory, or more correctly, in the form of image-memories or memory proper, which are the most direct and disinvested forms of access to the past. These memories are the condition of perception in the same way that the past, for Bergson, is a condition of the present. Whereas the past in itself is powerless, if it can link up to a present perception it has a chance to be mobilized in the course of another perception's impulse to action. In this sense, the present is not purely in itself, or self-contained; it straddles both past and present, requiring the past as its precondition, and is oriented toward the immediate future.[26] Both perception and action are oriented to the present as the threshold of the future. The present consists in the consciousness I have of my body and my body's capacity to harness the object in action. Memory, the past, has no special link with or proximity to my body. It has no interest in mobilization of bodily energies or forces; rather, it enters consciousness at its most pure only in a state of relaxation, where the interests of the present are diminished or minimized, where speculation and reflection take over from the pragmatic impulse to action.

Bergson asserts that the past and the present cannot be understood as mere differences of degree or magnitude, the past a receded or diminished version of the present that it once was. If this were the case, then we would have difficulty in distinguishing between a weak perception and a strong memory: they would form a continuum with no decisive break; there would be a point at which the one blurs into the other. If this were the case, we might, for example, mistake the perception of a weak sound for the memory of a loud one (*MM* 239). Yet we are never confused about whether a sensation is perceptual or remembered, even if we can be confused about whether the sensation is real or imagined. The essential difference between them is that the present acts on me and beckons me to act, whereas the past is powerless and inert, dependent on my stillness and detachment to gain some place in my present awareness.

What, then, distinguishes the past from the present? In what do their differences in kind consist? This can be answered only by specifying the privileged role and nature of the present. Both past and present are modalities or dimensions of duration: the past is time already passed, and the present is "the instant in which it goes by" (*MM* 137). But how long is this "instant"? How are we to measure its boundaries? When does it cease to be present? When we intellectualize about the present, we would like to see it as an instant, clear-cut and self-contained—what Bergson calls "an ideal present." The living present, that which we concretely experience, has its own duration; it has no minimal units, no instants or length, except those imposed retroactively through analysis.

My present extends itself to include those memories of previous "instants" that still generate sensations and cannot, except arbitrarily, be cut off from the present; it also includes the motor schemas that prepare us for action in the next instant. It is in this sense that "the present is sensori-motor" (*MM* 138). It extends itself to include sensory inputs of previous impressions and also potential motor outputs or schemas of action that anticipate the future. The present, then, is not an instant, a measurable and regulated moment; it is a dynamic concept that extends itself to include the fringes that touch both past and present. The length of the present varies according to the continuity that it assumes, the duration it occupies. For example, if we are to understand a statement, we must include as part of the present not only the word we are currently reading or hearing but all of those words, phrases, sentences, and so on that we need for understanding the present word. The present for a listener or reader is extended as far as he or she chooses, potentially including the whole text, and its context of predecessors and heirs. In the case of a melody (many of Bergson's most striking examples are musical), the present does not correspond to a note, because the note itself has a variable duration. The present must be capable of containing the whole of the melody, not just any notes that compose it. The present is that synthesis of all the undivided elements that constitute its continuity.

The present must be understood as elastic, capable of expanding itself to include what from the past and immediate future it requires to remain in continuity with itself, to complete its present action. It has no measurable length, for it takes as long as it takes to perform a continuous action; the present may be nearly instantaneous for a quick action (the blink of an eye), or it may stretch itself to include minutes, hours, days, and even longer. When, for example, we talk of geological or evolutionary duration, we may define the present in terms of centuries or even millennia.

The present, defined as it is by perception and action, is fundamentally, and paradoxically, linked to space. The distance of an object in space is a direct measure of the threat or promise of that object in time: the nearer the object, the more immediate its impact on the perceiver. Space signifies or represents our near future, that future which is already tied to the present, that future which is implied in or posited by our current perceptions and actions. Space, perception, objects, action are all aligned through my body's location and placement as an object among the other objects in the world.

As we will see later in this chapter, the present contains not only what is active in real terms; it also musters the virtual action that encompasses past and future in a continuous movement. The past is that which is exhausted in its real activities, and which can derive vitality or enervation only from the force of the present; its influence is now virtual, and can be reenergized only by being placed on a path that links it to actualization, to making, to movement and the future. If the past and present can be understood in operational terms, the present as that which is active and the past as that which can no longer act except through the borrowed energy of the present, then the past is not merely psychological but also ontological. It exists, whether we remember it or not, and it exerts whatever is unexhausted in it only through access to the present.[27]

This is indeed the primary political relevance of the past: it is that which can be more or less endlessly revived, dynamized, revivified precisely because the present is unable to actualize all that is virtual in it. The past is not only the past of *this* present, but the past of every present, including that which the future will deliver. It is the inexhaustible condition not just of an affirmation of the present but also of its criticism and transformation. Politics is nothing but the attempt to reactivate that potential, or virtual, of the past so that a divergence or differentiation from the present is possible. Bergson is one of the few theorists to affirm the continual dynamism, not of the present, but of the past, its endless capacity for reviving and regenerating itself in an unknown and unpredictable future.

The past cannot be identified with the memory-images that serve to represent or make it actual for or useful to us; rather, it is the seed that can actualize itself in a memory. Memory is the present's mode of access to the past. The past is preserved in time, and the memory-image, one of its images or elements, can be selected according to present interests. Just as perception leads me to objects where they are, outside of myself and in space, and just as I perceive affection where it arises, in my body (*MM* 57), so too, I recall or remember only by placing myself in the realm of the past, where memory subsists. Memory, our mode of access to the past, is thus, paradoxically, not *in us*, just as perception is not *in us*. Perception takes us outside ourselves, to where objects are (in space); memory takes us to where the past is (in duration). And incidentally, language, too, takes us to where concepts are. In each case, this movement—in space, in time, in concepts—is possible only because of our capacity to (temporarily, or with some effort) disconnect from our immersion in a tensile and expanding present to undertake the leap that these movements outside ourselves, and outside our habitual behavioral schemas, require. The past is not accessible to us as if it were stored or recorded in a file or document; we do not simply seek for the place in which a memory resides and find the past in all its detail there. This is both to spatialize duration and to treat memory as if it were the perception of a thing. Bergson often talks of the act of disconnection that must occur for us to access the virtual, the past, or other languages: "Whenever we are trying to recover a recollection, to call up some period of our history, we become conscious of an act *sui generis* by which we detach ourselves from the present in order to replace ourselves, first, in the past in general, then in a certain region of the past—a work of adjustment, something like focusing a camera. But our recollection still remains virtual; we simply prepare ourselves to receive it by adopting the appropriate attitude. Little by little it comes into view like a condensing cloud; from the virtual state it passes to the actual; and as its outlines become more distinct and its surfaces take on color, it tends to imitate perception" (*MM* 133–134).

Bergson argues that the past would be inaccessible to us altogether if we could gain access to it only through the present and its passing. The only access we have to the past is through a leap into virtuality, through a disconnection from the present and a move into the past itself, seeing the past is outside us and we are in it rather than its being located in us. The past exists, but it is in a state of latency or virtuality, as the potential of other ongoing presents. We must place ourselves in it if we are to have recollections, memory-images, and this we do in two movements or phases. First, we place ourselves into the past in general (which can occur only through a certain detachment from the immediacy of the present), into the ontological milieu of duration itself; then we place ourselves in a particular region of the past. If our seeking a particular memory-image is unsuccessful, we must return to the present and seek again.

In the first movement, through a detachment of our attention from the present, we must "at a stroke" leap into the past in general, an ontological element different in nature from the present and its tendency to spatialization. This is the preparatory gesture that readies us for specific recollections.[28] Then, in a second movement, we seek our way in the past in order to locate a specific memory-image, like focusing the lens of a camera to more sharply represent the object. The first movement is a leap (not unlike the first movement of Nietzsche's dice throw: the throw of the dice upward) into the virtual order of the past; the second (like the fall of the dice back to earth) is a particular attunement to the details of memory (or, for example, a statement in another language, or a melody being performed), a relative passivity that enables us to take in, to actualize, the past in a concrete memory (or a concrete act of linguistic understanding), that is, in a current affection that prepares itself for future motor schematization. We can find the memory for which we search only by placing ourselves in the past itself, where the past may materialize itself in a memory, psychologize itself, connect itself to the body.

It is a similar movement that reveals language to us, for our ability to understand is necessarily bound by our access to memory. (Language also functions as virtual: utterances, concrete statements, particular discourses actualize or concretize a language that is itself pure and unlimited, virtual potential, the potential to say anything.) We must place ourselves in a position to be receptive to what we hear or read, through an act of psychical and motor preparation. We do not understand through the accretion of individual words, creating the meaning of a statement step by step as each word is articulated. We develop an understanding of language all at once. We can understand language only through a wholesale immersion in its conceptuality or sense, and only from there can we materialize the sounds we hear or the words we read into a form of comprehension. The same is true, Bergson claims, of a mathematical calculation: we can conceive it only by doing it, mentally or physically repeating its steps, by first placing ourselves in its ontological terrain or plane, within the frame and terms of its own conceptuality.

Consider the example of hearing someone else speak: the listener must place himself or herself in the conceptual orbit of the speaker, through a leap into meaning-receptivity in general, and then focus on a particular articulation. The initial leap into conceptuality is impossible if one does not understand a language. As Bergson makes clear, we cannot even discern the distinction between words or significant sounds in a language whose conceptuality we cannot enter; it is only from within its conceptual or sense-bearing frame that its units can be discerned and its intelligibility deciphered. For someone who does not understand a language, what passes between two speakers is nothing but indecipherable noise: it emerges as language, as meaningful articulation, only to the degree that someone, even if he or she cannot speak, can immerse himself or herself in its signifying order (see *MM* 109).[29]

Bergson conceives of the past in terms of a series of planes or segments, each one representing the whole of the past in a more or less contracted form. The present can be understood on such a model as an infinitely contracted moment of the past, the point where the past intersects most directly with the body. It is for this reason that the present is able to pass. He represents this diagrammatically in the famous cone figure (figure 2).

The cone SAB represents the totality of memory, in its different degrees of contraction or relaxation. The base AB is situated in the pure past and is

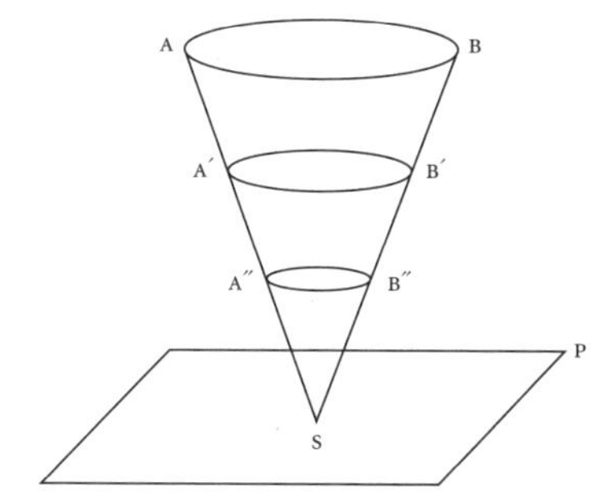

Figure 2
Source: Henri Bergson, *Matter and Memory*, trans. N. M. Paul and W. S. Palmer (New York: Zone, 1988), 162, fig. 5.

unable to link with the present. The point S indicates my continuing, mobile present. The plane P is my actual present representation of the universe, my point of direct contact with objects. S is the locus of the sensorimotor functions, the point at which memory is the closest to action, where it is the most compressed and the most connected to the present. Bergson suggests that automatic or simple animal existence is focused on this point, where memory is more or less useless and where habit and the automatic and mechanical synthesis of the past are the most action-oriented. The segments AB, A′B′, A″B″ are repetitions of memory in increasingly dilated form. The more expansive and detailed, the less accessible is memory to present action, the less relevant it is to the present, and the more it is capable of occupying those states of greatest relaxation, sleep and dream. The most dilated level thus represents a dream-plane, the most languid and expansive of all memories, where memories can elaborate themselves for their own sake instead of being subordinated to a current interest.[30]

Each segment has its own features, although each contains within itself the whole of the past. Memories drawn from various strata may be clustered around idiosyncratic points, "shining points of memory," which are multiplied to the extent that memory is dilated (*MM* 171). Depending on the recollection we are seeking, we must jump in at a particular segment; to move on to another, we must, as it were, return to the present, to the point S, and make the effort of another leap.

This diagram could equally represent language or conceptuality as it does the past: each involves the production of different orders of intensity, the necessity of mixtures or combination of mind (signifieds, ideas, solutions) and matter (signifiers, numerals, problems) and the necessity of a temporary detachment from the force or immediacy of the present. It is thus only through a similar structure of compression and dilation that we can detach ourselves from the present to understand linguistic utterances or make conceptual linkages.[31] In all three cases, this leap (the very leap constitutive of radical politics, as we will see) involves landing in different concentrations of the past, language, or thought, which nonetheless contain the whole within each of them in different degrees.

Bergson's understanding of this cone of the past raises a number of complications, at least for conventionalized representations of time that tend to depict it on the model of space. (Ironically, of course, any diagram spatializes the temporal, presents succession through simultaneity.) When we understand the past as a faded present; when we see the present as an illuminating spotlight on events for their brief moment of existence, which then casts them into shadows of obscurity; when we see the present as the linear completion of the past, we regard the past and the present as mere differences of degree. We spatialize time. We are unable to understand how the past coexists with the present, the ways that time is rendered paradoxical. Space represents relations of contiguity and coexistence, which include relations of containment. In duration, by contrast, relations of succession function to frame relations of simultaneity,[32] and no "object" can be isolated from another or function to include or contain another.

The present could never be present without the weight of the past which it carries with it in an ever-accumulating entwinement: the cone grows with each "moment," though the present remains always located at the intersection of the most contracted forms of memory with motor actions. Each moment carries a virtual past with it: each present must, as it were, pass through the whole of the past. This is what is meant by the past in general: the past does not come after the present has ceased to be, nor does the present become, or somehow move into, the past. Rather, it is the past which is the condition of the present; it is only through its preexistence that the present can come to be. Bergson does not want to deny that succession takes place; of course, one present replaces another, but such real or actual succession can take place only because of a virtual coexistence of the past and the present, the virtual coexistence of all of the past with each moment of the present. This means that there must be a relation of repetition between each segment, whereby each segment or degree of contraction/dilation is a virtual repetition of the others, not identical, but a version. The degrees of contraction or dilation that differentiate segments constitute modes of repetition in difference.[33] But perhaps what is most significant about the ballast of the past is that it is not only the accompanying condition for every present, but also the (virtual) condition for any and every future, although the future remains unconnected by any direct means to the force of the present. Its resources come from that which in the past is unconsumed by the present.

To briefly summarize our understanding of Bergson's account thus far:

1. Duration must always be regarded as a continuity, a singular whole. When duration is divided, which fundamentally transforms its nature, it can be regarded as time, the scientific, measurable counterpart of space; but in itself, and not subordinated to the exigencies of practical and scientific action, it is indivisible, continuous, inscribed by movement, always a whole.

2. Duration is both singular and a multiplicity. Each duration, each movement, each act forms a continuity, a single, indivisible whole; and yet, there are many simultaneous durations, as many perhaps as there are actions, which implies that all durations participate in or can be linked through a generalized or cosmological duration, which allows them to be described as simultaneous. Duration is the very condition of (the spatial characteristic of) simultaneity, as well as succession. An event occurs only once: it has its own characteristics, which will never occur again, even in repetition. But it occurs alongside, simultaneous with, many other events, whose rhythms are also specific and unique. Duration is thus the milieu of qualitative difference, and each difference it proliferates is different in kind, unique in itself.

3. The division of duration—which occurs whenever time is conceptualized as a line, counted, divided into before and after, made the object of the numerical, rendering its analogue continuity into digital or discrete units—transforms its nature, that is to say, reduces it to modes of spatiality. If, as Bergson suggests, space is the field of quantitative differences, of differences of degree, then the counting of time, its linear representation, reduces and extinguishes its differences of kind to replace them with differences of degree (the source of many philosophical illusions and paradoxes, most notably Zeno's paradoxes).

4. One of the most significant differences of kind within duration (which is commonly misunderstood as a difference of degree) is the distinction between past and present. The past and the present are not two modalities of the present, the past a receded or former present, a present that has moved out of the limelight. Rather, the past and the present fundamentally coexist; they function in simultaneity. Bergson suggests that the whole of the past is contained, in contracted form, in each moment of the present. The past lives *in time*. The past could never exist if it did not coexist with the present of which it is the past, and thus of every present. The past would be inaccessible to us altogether if we could gain access to it only through the present and its

passing. The only access we have to the past is through a leap, through a move into the past itself, given that, for Bergson, the past is outside us and that we are in it. The past exists, but it is in a state of latency or virtuality.

5. If the present is the actuality whose existence is engendered by the virtual past, then the future remains that dimension or modality of time that has no actuality either. The future, too, remains virtual, uncontained by the present but prefigured, rendered potential, through and by the past. The future is that over which the past and present have no control: the future is that openness of becoming that enables divergence from what exists. This means that, rather than the past's exerting a deterministic force over the future, the future is that which overwrites or restructures the virtual that is the past: the past is the condition of every future; the future that emerges is only one of the lines of virtuality from the past. The past is the condition for infinite futures, and duration is that flow that connects the future to the past that gave it impetus.

8. THE PHILOSOPHY OF LIFE

In his earlier work, Bergson claims that a distinction between subjective and objective (consciousness and matter, or duration and spatiality) can be formulated in terms of the distinction between the virtual and the actual. Objects, space, the world of inert matter, are entirely actual; they contain no elements of the virtual. Although matter may well exceed the images that we have of it, although there is more in matter than in our images of it (matter itself being myriad images, only some of which interest us), the object is nonetheless of the same kind as our images; that is, because matter has no virtuality, no hidden latency, it is largely assimilable to the images we have of it: "There is in matter something more than, but not something different from, that which is actually given. Undoubtedly, conscious perception does not compass the whole of matter, since it consists, in as far as it is conscious, in the separation or 'discernment,' of that which, in matter, interests our various needs. But between this perception of matter and matter itself there is but a difference of degree and not of kind, pure perception standing toward matter in the relation of the part to the whole. This amounts to saying that matter cannot exercise powers of any kind other than those which we perceive. It has no mysterious virtue; it can conceal none" (*MM* 71).

The Actual and the Virtual

If everything about matter is real, if, that is, matter has no virtuality, the material object's medium is spatial. The object, though it is subject to change, does not reveal more of itself in time: it is "no more than what it presents to us at any given moment" (*MM* 68). What duration, memory, consciousness bring to the world is the possibility of an unfolding, a narrative, a hesitation. Not every thing is presented all at once. Matter is an enfolding that consciousness is capable of unfolding.[1] This is what life (or consciousness) brings to the world: the remembrance of the past, the history submerged or lying behind the present, whose resources are not completely depleted for they can reinvigorate the present and help generate the new, which, for Bergson, is precisely the movement of the actualization of the virtual. Duration is the subsistence of the past in the present and the capacity of this reimbued present to generate an unexpected future beyond that of imminent action. If the immediate future, the next "moment," is already contained in the present, then the future itself must be radically discontinuous with the present, a surprise, new: "Thus the living being essentially has duration; it has duration precisely because it is continuously elaborating what is new and because there is no elaboration without searching, no searching without groping. Time is this very hesitation" (1992, *The Creative Mind* [hereafter CM], 93).

If matter can be placed on the side of the actual and the real, and if mind, life, or duration is to be placed on the side of the virtual, we need to be more clear about how Bergson distinguishes the two oppositions, between the virtual and the actual, and the possible and the real, and how he aims to problematize and displace the possible/real relation with the virtual/actual relation. This relation has come to generate considerable contemporary interest, especially from those concerned with the question of politics.[2] This preference for the terms that generate the unexpected rather than the planned or prepared for will prove of major political significance for feminists, antiracists, and political activists of all kinds. Whether they recognize it or not, the aim of all radical politics is the production of a future that actively transforms the dynamics of the present, and this may involve precisely an unpredictable leap into virtuality—into both senses of the virtual: the virtual past and the virtual future—which carries no pregiven plan or guarantee except a derangement of the present order, a movement of rendering its order insecure and replaceable. This leap into the virtual is always a leap into the unexpected, which cannot be directly planned for or anticipated, though it is clear that it can be prepared for. The resources for this derangement, as Nietzsche recognized, come only from a judicious, or wild, return to the past, or at least to that part of the past that has not been directly utilized, used up, by the present. Politics is this untimely activation of the virtuality of the past as challenge to the actuality of the present.

Bergson recognizes the activity of closure that regulates the process of realization. If something is possible before it is real—say, the rules of etiquette, the composition of a musical piece, the creation of a work of art or literature, a scientific theorem—a belief we commonly assume in everyday life, we assume that it is already there waiting for discovery rather than being invented. We understand preliminary sketches, drawings, notes as the possible early version of a completed work. Yet Bergson argues that it is an illusion to understand the possible as the preexistence of the real. Did the laws of geometry exist before the ancient Greeks discovered them? Was the *Mona Lisa* possible before Leonardo painted it? Bergson argues that no work is possible before it is real. Imagine as we may the contours and details of an event, an object, an activity in advance, we are always surprised by its actual characteristics. A work of art, or a body of knowledge, is not possible before it is real. At best, after it is real, after it is created and exists, we can say that it "will have been possible" (*CM* 118).

The process of realization, of making the preexisting possible into a real, is governed by two principles: resemblance and limitation. The real exists in a relation of resemblance to the possible. Indeed, the real is an exact image of the possible, with the addition of the category of existence or "reality." Conceptually, in other words, the real and the possible are identical (because, as Kant argued, existence is not a quality or attribute). Moreover, the process of realization involves limitation, the narrowing down of possibilities, so that some possibles are rejected and others made real. The field of the possible is thus wider than the real. Implicit in this pairing is a preformism: the real is already preformed in the possible insofar as the real resembles the possible. The possible passes into the real through limitation, the culling of other possibilities. But through this resemblance and limitation, the real comes to be seen as *given* rather than *made*, as an inevitable outcome, merely waiting for real existence.

On such a model, the possible is both more than but also less than the real. It is more insofar as the real selects from a number of possibles, limiting their profusion; but it is less insofar as it is the real minus existence. Realization is a process in which creativity and production have no place. Making the possible real is simply giving it existence without adding to or modifying its conception. It is only materializing that which is already perfect as idea. Does the possible produce the real, as our everyday beliefs imply, or does the real in fact project itself backward to produce the possible as its retrospective shadow? Bergson argues that it is only when the real exists as such that we can understand that it must have been possible. The possible is thus not the prefiguring of the real, a rehearsal before its actualization; it is instead the reassurance that the real gives itself that it was inevitable, that it was preordained, already given.

Bergson claims that the processes of resemblance and limitation are subject to the philosophical illusion that consists in the belief that there is *less* in the idea of the empty than the full and *less* in the concept of disorder than order, when in fact the ideas of nothing and disorder are *more* complicated than of existence and order:[3]

> I find the same illusion in the case in point. Underlying the doctrines which disregard the radical novelty of each moment of evolution there are many misunderstandings, many errors. But there is especially the idea that the possible is *less* than the real, and that, for this reason, the possibility of things precedes their existence. They would thus be capable of representation beforehand; they could be thought of before being realized. But it is the reverse that is true . . . We find that there is more and not less in the possibility of each of the successive states than in their reality. For the possible is only the real with the addition of an act of mind which throws its image back into the past, once it has been enacted. But that is what our intellectual habits prevent us from seeing. (CM 99–100)

In this sense, then, the real creates an image of itself, which, by projecting itself back into the past, gives it the status of the possible. The ideally preexistent, where existence precedes matter, precludes any notion of surprise, invention, newness: it minimizes the riskiness, the struggle, the precariousness of invention, of the production of something new, whether a text, a work of art, or a theorem, giving each the aura of the expected and the assimilable. It justifies what is real now as the best, the strongest, the most justified of the options the possible provided us. To reduce the possible to a preexistent phantom-like real is in effect to curtail thinking the new, of thinking an open future not bound to the present:

> As reality is created as something unforeseeable and new, its image is reflected behind it into the indefinite past; thus it finds that it has from all time been possible, but it is at this precise moment that it begins to have been always possible, and that is why I said that its possibility, which does not precede its reality, will have preceded it once the reality has appeared. The possible is therefore the mirage of the present in the past; and as we know the future will finally constitute a present and the mirage effect is continually being produced, we are convinced that the image of tomorrow

> is already contained in our actual present, which will be the past of tomorrow, although we did not manage to grasp it. (101)

The possible must be in excess of this restrictive understanding, for it makes both a reality that corresponds to it possible, and also makes possible the projecting back of the present reality into a modality of the past.[4] A different, more positive understanding of the productivity of invention is required to add the dimension of the uncertainty and struggle required to actualize, to bring something into existence or modify its operations than this otherwise smooth, almost inevitable transition from the possible to the real implies.

Bergson aims to replace the notion of realization with that of actualization. Actualization involves a process of genuine creativity and innovation. His point, though, is not a semantic one—the claim that one should replace the terms "possible" and "real" with the terms "virtual" and "actual"—but, rather, conceptual. We need to think the relations between the past and the present not in terms of the possible and the real, but in terms of the virtual and the actual.

The virtual/actual relation is governed by the two principles of difference and creation. For the virtual to become actual, it must create the conditions for actualization: the actual in no way resembles the virtual. It is produced through a mode of differentiation from the virtual, a mode of productive or creative divergence. The lines of actualization are divergent, heterogeneous, the emanation of a multiplicity from a virtual unity, creating the varieties that constitute creative evolution. This is a movement of the divergent paths of development in different unpredictable and unforeseeable series and directions.

The movement from a virtual unity to an actual multiplicity requires that there is a certain leap, this time a leap of innovation or creativity, the surprise that the virtual leaves within the actual. If realization is the concretization of a preexistent plan, program, or blueprint, by contrast, actualization is the opening up of the virtual to what befalls it. It is fundamentally unpredictable, innovative. In the terms of another discourse, actualization is individuation, the creation of singularity (whether physical, psychical, or social), insofar as the processes of individuation predate the individual yet the individual is a somehow open-ended consequence of these processes. Individuation contains the "ingredients" of individuality without in any way planning or preparing for it.[5] This indeed is what life, élan vital, is of necessity: a movement of differentiation of virtualities in light of the contingencies that befall it.[6]

Insofar as time, history, change, the future need to be reviewed in light of this Bergsonian disordering, perhaps the openendedness of the concept of the virtual may prove central in reinvigorating a politics embracing the future by refusing to tie it to the realization of possibilities (the following of a plan, a preset political agenda, a set of political values or goals given in advance or in universal terms) and linking it to the unpredictable, uncertain actualization of virtualities (a politics directed to means more than ends, directed to unorganized eruptions more than a given platform, to upheavals and rearrangements in social and political life rather than to its concerted reorganization). Feminist, queer, antiracist, and postcolonial struggles find themselves divided in these two directions, which can also be articulated in terms of the distinction between egalitarian struggles (which have a determinate goal of the more equal distribution of given resources) and those directed to the exploration of difference (which have indeterminable goals, indefinite aims, unclear aspirations). Bergson's understanding of the unpredictable impetus of the virtual, its potential for eruption and change in the present and future, may serve as a surprising source for feminist and antiracist conceptions of struggle and transformation.

Science and Intelligence

Matter and Memory elaborates Bergson's understanding of the distinction between space and duration, the actual and the virtual, the material and the conscious, extensity and the intensive, differences of degree and differences in kind. These pairs of terms—themselves fundamentally different in kind, as we have seen—are differentiated through their relations to practice: the first term in each of these pairs is manifest, measurable, ascertainable through perception or its technological augmentation, and is the object of potential action; the second term is latent, invisible, subjective, accessible only through direct experience (which, as we will see in the next chapter, is capable of being developed into the rigorous philosophical method of intuition) and thus incapable of scientific or objective verification. The sciences, especially physics and chemistry, have been remarkably successful in enhancing our access to and control over the first of these terms but, Bergson argues, have been spectacularly unsuccessful in penetrating the mysteries of biological or psychical life or understanding its nonenumerable elements and their integration.

At no point does Bergson criticize or renounce the value of science and its power to help manipulate, regulate, and anticipate or predict matter. The natural sciences, which operate primarily through resort to formulae, functions, algorithms, that is, through congealing what is fluid into what is solid, identical, measurable through the creation of minimal units and the conceptualization of their relations within a closed or regulated system, should not overstep the boundaries of their relevance. Whenever we attempt to scientifically measure, contain, regulate, and predict life and its durational complexities—social, historical, and psychological as well as biological—using the privileged techniques of physics and chemistry (decomposing and recomposing, analytic and synthetic), we lose something essential about our object of investigation; we reduce the second term in each of the pairs to the first. We impose on living forms the rigid structures of matter, and the techniques that make matter accessible in some ways change, reduce matter to our needs. This imposition is not without some relevance, insofar as living beings are also and inextricably material objects, but the most essential characteristics of life—its duration, its temporality, its span of living, its qualitative intensity—disappear in the process.

A recurrent theme in all of Bergson's writings is the inadequacy of intelligence, science, and thus epistemology to address the most central questions of existence. Science has remained impressively adept in the manipulation of matter, but remarkably inept, mechanical, and abstract in understanding what is living and subject to change. Both his earliest critics, who affirm the more or less limitless value of science and logic in explaining and ordering the real (Russell, Huxley), and his postmodern opponents, who affirm that the real is only ever accessible through synthesis (Derrida and deconstructionism, common characterizations of postmodernism), share a belief in the coincidence of epistemology/representation/discourse with the real, the collapse of ontology into epistemology. If there is a real that cannot be measured, represented, or known, then it remains, so it is believed, below philosophy's threshold of relevance. What these otherwise antagonistic positions—positivism, postmodernism—share, it seems, is that representations, whether mathematical (for Russell) or discursive (for Derrida), are our only means of access to the real, the way we signify or construct the real. Neither considers, as Bergson suggests, that the real is not only that which discourse produces or depicts but also that which occasions and contextualizes action, including all the representational actions that discourse performs.

The discursive, the symbolic, and the formulaic do not exactly map or cover the contours of the real. At best, they represent certain regularities, certain relations extracted from the real. Bergson affirms, in contrast with this prevalent representationalism, the necessity that the real encompasses the known and the knowable. The real must be *larger* than and *outside* the known; it must contain not only the known but also what knowing leaves out, the interstices, intervals, gaps between and within knowable units, the differences covered over by apparent identity, the open rather than closed context of all material events, whose influence and effects reach "even to the stars" (1921, *Mind-Energy* [hereafter ME], 30). Knowing, forms of intellect, traditional models of science all privilege the spatial outline of material objects: they spatialize and thus cannot understand the particularity or difference of duration from space: "[Intellect] dislikes what is fluid, and solidifies everything it touches. We do not *think* real time. But we *live* it, because life transcends intellect" (1944, *Creative Evolution* [hereafter CE], 46).

Our intellect, fashioned by and an effect of evolution, functions primarily to prepare for immanent action. This is its evolutionary advantage: it gives intelligent creatures access to a wider range of actions than those provided by reflex or automatic actions, by instinct or through habit, and thus access to a wider range of possible innovations. The intellect functions primarily through *recognition*, through organizing material by what is already known. The intellect thus tends to submit the unknown to the principles of the known, not only to recognize but to precognize, anticipate in advance, what is to come. Moreover, the intellect functions primarily through its analytic capacity to divide up a continuum—potentially infinitely, or at will—into discrete elements or atoms, into simple components, a process that facilitates the movement of recognition, the reduction of the new to a recombination of the old. The intellect tends to remain insensitive to slight variations and changes, for it is through these gradual and often apparently insignificant changes that innovation and change occur in a continuous manner. The favored mode of intellectual activity is thus bound to repetition, to the static, to that which remains the same over time.[7]

Rather than the highest achievement of evolution, the intellect is thus always a step behind the real, that is, the dynamic force and impetus for matter and for the creation of life that evolutionary development makes possible. Life is always richer, more complex *and* more simple, more diverse *and* more unified than the intellect can comprehend. Science, including (perhaps especially) the biological sciences, is oriented by the intellect and the practical concerns to which it is implicitly directed, and is naturally drawn to the double reduction of continuity to atomic elements and of present features to past ones. But the operation of life itself, in its ever complexifying, ever creative immersion in duration, that is, in its evolutionary embeddedness, resists this process of scientific reduction, which can, at best, gain only a partial view of materiality in its interlocking integration or in life's always particular and unexpected emergence and unfolding.

Bergson's position is not that of the skeptic, for whom knowledge has no value or its value needs to be proven. Rather, for him, there are a number of different kinds of knowledge. The knowledge provided by the intellect, which science refines and elaborates, finds itself most at home with material objects, with inanimate things, "especially solids, where our action finds its fulcrum and our industry its tools" (CE ix). It is a knowledge that finds its natural base in geometry, the intellect's propensity for abstract or diagrammatic spatial representation of perfect relations between homogeneous and measurable solids, which reveals the "kinship of logical thought with unorganized matter" (ix), the affinity of the procedures of mind with the raw operations of matter. Science finds its proper place as the extension of our bodily action, the most rational means by which we organize and understand matter to serve our needs.[8] Science, in his conception, is thus always tinged with a technological or engineering orientation: its inherent goal is the facilitation of more and more complex and far-ranging action.

What scientific, logical, and mathematical forms of knowledge must leave unconsidered are differences in kind, durational or qualitative differences, which they reduce to differences of degree in order to calculate and recompose them, leaving out the connections between objects, the processes by which objects blend with each other and transform each other. In other words, the processes essential to understanding the integrity, cohesion, and variability of life are lost. Intellect cannot encompass the life that produced it. When the intellect directs itself to life in its various permutations, it transposes the categories and methods that have successfully enabled much of the inorganic world to be deciphered and predicted onto that which has no principle of identity, which has no clear-cut divisions and boundaries: "Who can say where individuality begins and ends, whether the living being is one or many, whether it is the cells which associate themselves into the organism or the organism which dissociates itself into cells?" (CE x).[9]

Continuity and Change

Most of Bergson's efforts in his most renowned text, *Creative Evolution*, are directed to freeing up the study of life from the interventionist techniques of the natural sciences, most particularly to enabling the Darwinian understanding of evolutionary becoming to rid itself of the overlay of scientism that has either reduced it to a form of mechanism or automatism, or to teleology, a position he also describes as finalism. Darwin's work on the organic elaboration of life is the heart of *Creative Evolution*: Bergson aims to purify (neo-)Darwinism of the incursions of the predictive sciences, but also to show Darwin's relevance to the evolution of all aspects of life, to social, mental, artistic, and conceptual evolution, which he understands as creative evolution. Darwin's own insights may help to restore to philosophy as well as science some understanding of the fundamental mobility of life, its position in the flow of duration that will enable these knowledges to develop beyond their current forms. We need to return knowledge to the stream of life from which it is drawn, which means producing knowledges that are somehow more in tune with the real articulations of the world, its real differences, its qualitative intensities, knowledges that are capable of making themselves particular to suit their objects, knowledges in which "no quality, no aspect of the real would be substituted for the rest ostensibly to explain it" (CM 158), knowledges that are able to accommodate duration, change, and transformation. Bergson remains resolutely Darwinian: in Darwin's own writings he finds the first scientific intimations of the irreducibility of time to space and of the past to the present, and the subsistence of time in space and the past in the present.

Bergson seeks a model of change that does not subordinate it to the things that thereby change, a model of mobility that does not submit it to the privilege of immobile states or elements or recompose it from the addition of successive immobilities, as Zeno had sought to do in his various paradoxes of motion. Bergson seeks change in itself, change that does not simply ripple over unchanging things, change that is not carried along by things but that problematizes the very stability of what counts as an immobility, a thing, that which is cut out from its environment. We feel more at ease, more capable of being in control of our material circumstances, when we deal with stable objects. But this stability is itself the coordination, the simultaneity or synchronization of change, like two trains moving on parallel tracks at the same speed, rather than its elimination. The ever-changing movements that constitute my subjectivity naturally function in parallel with those ever-oscillating frequencies or differences, of light, sound, and material connection that constitute the material object, enabling me to act on and with it.[10] It is because my ever-changing continuity is (partially) synchronous with the oscillations of matter that I can perceive objects, even though my perception is clearly incapable of access to all of the differentiations and oscillations that make up the relative stabilization of objects. There is no immobile thing, state, moment underneath mobilities, supporting or enabling them: "There are changes, but there are underneath the change no things which change; change has no need of a support. There are movements, but there is no inert or invariable object which moves: movement does not imply a mobile" (CM 173).

Change cannot be construed through objects, states, or qualities that are themselves unchanging: it cannot be reduced to the mode of passage from one state or condition to another, the state in itself remaining the same. A change between states is a change *of* states; or rather, between a change of state and a change in states there is only a difference of degree: "This amounts to saying that there is no essential difference between passing from one state or another, and persisting in the same state. If the state which 'remains the same' is more varied than we think, on the other hand, passing from one state to another resembles, more than we imagine, a single state being prolonged; the transition is continuous" (CE 2).

States, especially psychical states and processes, cannot be understood as distinct, self-contained things. They flow into each other continuously, imperceptibly, more as fugitive emanations, as processes, than as atoms or units. States merge one into the other, with no clear-cut separation between one and the other; their transition is indiscernible. If we artificially arrest this indiscernible transition, we can understand states as separate entities, linked by succession, but we lose whatever it is that flows in change, we lose duration itself. Duration is the survival of a previous state in a successive one, the merger or union of one state with another that cannot be separated from it, rather than the replacement of the previous state by its successor. The past is already contained within the present, not because the present is larger than the past but because the past is preserved in itself and the present is its culmination.

Because the past survives, because the previous state affects and is internally linked to successive states, duration does not admit of repetition. Even in the material universe, the exact cosmological conditions, contra Nietzsche, are unlikely to recur in their precise details, and thus repetition is only formal or nominal, usually a form of similarity or resemblance, a repetition in some respects rather than an identity of recurring states. Consciousness cannot repeat the same state, for, as a being with memory for whom the past persists into the present and is carried along with it at each moment, even an apparently exact repetition cannot erase the memory of the first occurrence, which means its repetition cannot be guaranteed to generate the same effects or results. The "same" state now affects a different person at a different historical moment that is already marked by the first event, for, as conscious and living beings, we carry our past with us and it affects, though it does not control, our actions and reactions to events: "Our personality, which is being built up each instant with its accumulated experience, changes without ceasing. By changing, it prevents any state, although superficially identical with another, from ever repeating it in its very depth. That is why our duration is irreversible. We could not live over again a single moment, for we should have to begin by effacing the memory of all that had followed" (CE 5–6).

As living beings, we *are* the accumulation and concretion of our history, of what has happened to us and what we have done, perhaps even before any personal or subjective existence. The past, including one's personal life, the past of one's parents, one's cultural history and even biology, are carried with every living being. The history of all living beings is contained not only in its full detail as world history, the past; it is also contained within all beings, compressed in their genetic lineage, in the living remnants of earlier times, their continuing inheritance from their earliest ancestors. Although we think and perceive with only the most immediate layer of the past as it straddles the present, we always act with the whole of our past, which is, in a sense, our "identity," our "personality," the only stability that is possible in living organisms. This past does not determine our present action (indeed, this is the mistake committed by both determinism and its opposite, free will: acts are caused, but they are not caused by logically distinct and temporally separable causes, they are not chosen out of a plethora of equally possible alternatives), for our present actions spring directly from and in continuity with our past. The present is not the repetition or completion of the past but its prolongation. The future cannot be contained in the present (not without reducing the future to the present) because it is an expansion or elaboration of the present rather than its distillation, essence, or inevitable consequence.

In the case of inorganic objects, however, the past is irrelevant and the future is, in a sense, already contained within the present arrangements and conditions of the object in its relations with other objects. Without a knowledge of the object's reciprocal influence over other objects, and the material world as a whole, prediction is possible only within artificially closed systems. The isolated object is not subject to duration. What is important here is the configuration of the parts of an object in space, rather than the relations of continuity and succession that regulate it over time. Spatial organization is in principle always repeatable, even if only through the intervention of external forces: "A group of elements which has gone through a state can therefore always find its way back to that state, if not by itself, at least by means of an external cause able to restore everything to its place. This amounts to saying that any state of the group may be repeated as often as desired, and consequently that the group does not grow old. It has no history" (CE 8).

Bergson's point is that we can understand or intellectualize objects and the world of matter only insofar as these are the means of our possible action. The material world is an integrated totality, and if understood as a whole, which is contrary to our need for practical action, it exhibits a duration of its own, a mode of its own unpredictability and novelty.[11] But when we isolate material systems, when we cut them out of the continuity in which they occur, we transform them and enable them to be schematized, outlined, rendered manipulable, to become the objects of scientific knowledge and predictability. Bergson thus distinguishes between our knowing the material world intellectually and our living in it bodily. We live the world in a manner much more complex and more integrated with matter than our conceptual apparatus is able to comprehend, largely because thought itself has a practical function: thought performs an abstract or schematized separation of objects from each other and of subjects from objects. It is in this sense that thought is inadequate to life, that life is always richer and more complex, more integrated and simple than thought can comprehend.

Duration is experienced most incontrovertibly in the phenomenon of waiting. Waiting is the subjective experience that perhaps best exemplifies the coexistence of a multiplicity of durations, durations both my own and outside of me, which may, by chance, coalesce to form a "convenient" rhythm or coincidence, or may delay me and make me wait. Bergson illustrates our intellectual tendency to spatialize time with the famous example of dissolving sugar:

> Though our reasoning on isolated systems may imply that their history, past, present and future, might be instantaneously unfurled like a fan, this history, in point of fact, unfolds itself gradually, as if it occupied a duration like our own. If I want to mix a glass of sugar and water, I must, willy nilly, wait until the sugar melts. This little fact is big with meaning. For here the time I have to wait is not that mathematical time which would apply equally well to the entire history of the material world, even if that history were to spread out instantaneously in space. It coincides with my impatience, that is to say, with a certain portion of my own duration, which I cannot protract or contract as I like. It is no longer something *thought*, it is something *lived*. It is no longer a relation, it is an absolute. What else can

> this mean than that the glass of water, the sugar, and the process of the sugar's melting in the water are abstractions, and that the Whole within which they have been cut out by my senses and understanding progresses, it may be in the manner of consciousness? (*CE* 9–10)

We must wait for sugar to dissolve: the temporality of matter, or at least that portion of materiality artificially isolated in the glass, is such that we must wait for its processes, we must submerge our desire to the temporality of matter's own durations. Their rhythms may or may not coincide with our own. The process of dissolving sugar takes a certain amount of time. It may be utterly predictable how long under particular conditions, and this time may be synchronized with and thus measurable by, say, a clock. My waiting consists in the failure of its dissolving to coincide with my desire. For the physicist, the time it takes to dissolve is given: everything would be the same if time could somehow speed up, like a cinematographic reel; the story would be the same whether speeded up or slowed down or, indeed, not played at all. But for the impatient coffee or tea drinker, this waiting is experienced as absolute: it is the amount of measurable time I must endure before I get what I seek. It is the failure of everything to be given at once, the necessity of one thing after another.

In his earlier works, *Time and Free Will* and *Matter and Memory*, Bergson understands matter as thoroughly inert, imbued only with spatial but not with durational properties. Matter, space, objects constitute a continuous present. They are, as it were, reborn anew at each moment, for they carry in present form what they need from the past while retaining no traces of the past as past. They are externally propelled into the future from the present as they were from the past. Time could be contracted or expanded, in principle, yet leave intact all the relations between terms. But in his later work, *Creative Evolution* and the contemporary essays he published in *Creative Mind* and *Mind-Energy*, space and duration, matter and mind can no longer be binarized, for they form a continuum, merge into each other, or rather, are divided into two tendencies, "descent" and "ascent," which roll and unroll, fold and unfold each other, insinuating each into the other with an inverse tension, in a continuity that resolves differences of kind into differences of degree.[12]

Matter in its totality participates in duration, and it is for this reason that life, too, must extend itself temporally. Not everything is given at once. Duration is that prolongation of events, processes, or objects into the future. But this does not apply to isolated physical systems, only to material systems in their complex interlocking totality. The world as a whole has duration, even if the systems our intellect isolates from each other within the world are regarded as amenable only to time, that is, to the spatialized or numerical reduction of duration. In the universe as a totality, there are no closed systems. The influence of each event, even at the most local level, may have ramifications and effects on many diverse events that may eventually reach to the furthest parts of the cosmos. In this sense, Bergson may be seen as a predecessor of contemporary complexity theory. This complex totality of materiality, this cosmological duration is that which coincides with our own duration whenever we are obliged to use, that is, isolate, the particular properties of matter for our purposes. We would not be able to utilize material objects and processes unless there is, at least at those points of overlap, a shared duration, a coincidence of the needs of life and the properties of matter. This coincidence occurs when we work with matter, but not when we conceptualize matter. Life introduces a force that, on the one hand, makes us coincide and work in rhythm with the duration(s) of the world and the objects and relations that constitute it and, on the other, enables us to conceptualize it and thus to separate ourselves from it: "There is no reason . . . why a duration, and so a form of existence like our own, should not be attributed to the systems that science isolates, provided such systems are reintegrated into the Whole. But they must be so reintegrated. The same is even more obviously true of objects cut out by our perception. The distinct outlines which we see in an object, and which give it its individuality, are only the design of a certain kind of influence that we might exert on a certain point of space: it is the plan of our eventual actions . . . Suppress this action . . . and the individuality of the body is re-absorbed in the universal interaction which, without doubt, is reality itself" (CE 11).

A being with no memory, no habits, instincts, or accretions of the past in the present, that is, which experiences no inner change, would share in the ever-renewed present of pure matter; such a being would blend in with the mechanical operations of the universe, would tend toward the inertia of matter itself, to automatism, to utter predictability.[13] Memory testifies to the persistence of the past, to the capacity of the past to persist through the present in the changes it wrought and to introduce disruption and surprise to the present.

Knowledge and Life

Darwin's model of the struggle between biological beings and their natural environment posits a series of ongoing and mutually transforming encounters between life and matter. Life accommodates itself ever more finely to the qualities and properties of matter and finds itself able to modify and develop these qualities and properties so that it not only transforms itself as an organic being, but also modifies and transforms its environment. For Darwin, neither life nor the environment in which the living find themselves is a static and unchanging entity: each is subjected both to inner transformation and to a dynamic interchange that perpetually modifies them both.

It is in *Creative Evolution*, whose very title marks a debt to Darwin's understanding of life, that Bergson explores in the most detail and depth his commitment to a Darwinian conception of life, and his profound modifications of Darwinism, especially of the social Darwinist and sociobiological tradition that followed.[14] In many ways, Bergson seemed to anticipate the increasing inhabitation of the sciences of life by the sciences of inorganic matter, the attempts of those who follow Darwin to render his work more and more precise through more detailed measurement and quantification. Instead of this trajectory, Bergson affirms a philosophical and ontological interest in Darwin's conception of the open-ended dynamism of life, its nonteleological orientation. Above all, he elaborates and develops Darwin's understanding of individual variation along the lines of Nietzsche's will to power. Instead of regarding evolutionary fitness as the passive adaptation to the active effects of the environment, Bergson sees life as an active, excessive, and inventive response to the provocation or stimulus of an environment, which induces in life, not just adequate or bare adaptation, but the capacity for immense—indeed excessive—development, elaboration, and complication: "While our [i.e., intelligence's] motto is *Exactly what is necessary*, nature's motto is *More than is necessary*,—too much of this, too much of that, too much of everything" (CM 249).

Emerging at a certain moment from chemical reactions whose consequences should have been predictable in principle but which operated as an unexpected eruption, life appears as a contrary current (ascent, enfolding, storage) to the distribution of energy in the nonorganic universe (descent, unfolding, expenditure). Whereas the rest of the universe winds down, expends energy, requires an external impetus to transform itself, living beings store and produce energy, internally modify themselves, producing a new dynamism in the universe, a force of self-proliferation.

Life emerges as an eruption of unpredictability in a material world that is otherwise structured by external forces or external relations between objects. This "current of life" enfolds, rolls up, stores for later and unexpected uses the energy that would have been expended automatically in inert objects. It injects a spark of novelty, invention, into what is otherwise predictable. This current is not the contradiction of the thermodynamic principles regulating matter (e.g., the conservation of energy, the tendency to entropy) but their complication and deviation, their passing into unexpected channels and pathways. In following the necessary or constitutive principles of matter, life also develops other goals, linked to self-elaboration, autopoesis, that inflect and delay those principles, or at least subordinate them to others (such as development and growth): "Living beings must be adapted to the conditions of the environment. Yet this necessity would seem to explain the arrest of life in various definite forms, rather than the movement which carries the organization ever higher. A very inferior organism is as well adapted as ours to the conditions of existence, judged by its success in maintaining its life: why then does life which has succeeded in adapting itself go on complicating itself, and complicating itself more and more dangerously? . . . Why has life gone on? Why—unless it be that there is an impulse driving it to take ever greater and greater risks towards its goal of an ever higher and higher efficiency" (*CM* 18–19).

He names this force that generates life always forward, that projects it to ever more complexity, the élan vital, a vital force or impetus. But this concept, the center of much criticism of Bergson, is misconstrued as a generalized supervening life force shared in common by all living beings. Bergson is careful to distinguish his position from that of vitalism, with which it is often confused.[15] Rather, it is a general name for the evolutionary impetus for increasing elaboration, differentiation, and specialization in living beings over the passage of large periods of time (many, many generations), that is, for the specific tendencies of change inherent in each particular form of life. Instead of mere passive chance generating variation, Bergson implies that there is a force or impetus of variation that makes life as varied yet unified as it can be, and leads life to vary or transform itself and its environment as much as it can, a force that is reminiscent of the Nietzschean conception of an inner will to power.

Bergson places himself, as in all of his arguments, in a position as it were midway between two otherwise opposed or polarized philosophical positions. In the case of matter and consciousness or memory, his focus on perception and action, as mixtures of difference, makes it clear that the polarities must meet and mix, must entwine themselves in each other as differences of degree. In the case of his analysis of evolutionary theory in *Creative Evolution*, he places himself in between the position he calls "radical mechanism," which understands life on the model of matter, reducing life to its smallest elements to understand the processes by which these elements can be reconstituted, and "radical finalism," or teleology, which explains processes and changes in terms of their outcomes or attainments. Neither, he argues, can understand change because, in spite of their otherwise antagonistic positions, each shares a belief that the processes of evolutionary change can be reduced to and explained by one or more of its elements: its causes, in the case of mechanism, or its ends, in the case of finalism, both of which are abstractions from the living whole that miss its essential characteristics, its durational becoming.[16]

Both mechanism and finalism attempt to contain the future in the present, in other words, to arrest that which is uncontained in the present through modes of knowledge that freeze time, that solidify matter, and that make life into a mere form of material object. Both are products of the intellect's tendency to think time through space, to make the ingredients of mobility immobile, and its inability to conceptualize duration as ceaseless change. Both mechanism (especially in its contemporary genetic form) and finalism leave out the evolutionary process itself. By this, Bergson means not only the evolutionary details of the emergence and development of a particular species or individual (this is what contemporary evolutionary theory has concentrated on with considerable success) but the more general movement of life itself, including not only those species that have succeeded, leaving a continuous path to the present, but also those that have grown extinct, that have become "terminal points in evolution" (*CE* 50).

This distinction between mechanism and finalism can be understood in other terms as the distinction between a preformism in which the virtual is reduced to the possible (the possible is the pregiven blueprint or plan of the real) or in which evolution is understood only in terms of actuals, one actual giving way directly to another without the mediation of the virtual (finalism) or where evolution is seen in terms of the impingement of the external influence of the environment and other contingencies (mechanism). If Monod, Dennett, Dawkins, and E. O. Wilson can be taken as representative of the mechanistic position, in which genetic forces, originary causes, alone are significant,[17] then perhaps Gould, Darcy Thompson, and others who focus on external form, on phenotype, may be seen as more or less contemporary versions of finalism.

Viewing the impersonal movement of life through all species as their procedures of variation or difference, Bergson does not see a single uniform thread of life (a single or universal élan vital) running through all forms of evolution. This would again affirm either a single origin or end to explain a multiplicity of processes; it would take either the cause or the effect of the process for the process itself. Instead, he affirms a plurality of movements, with no given aim or direction, which nonetheless all exhibit the force of time on the development of life: "Evolution has actually taken place through millions of individuals, on divergent lines, each ending at a crossing from which new paths radiate and so on, indefinitely . . . Roads may fork or by-ways be opened along which dissociated elements may evolve in an independent manner, but nevertheless in virtue of the primitive impetus of the whole the movement of the parts continues. Something of the whole, therefore, must abide in the parts" (CE 53–54).

The vital impulse is not a generic life force, but an initial impetus ("something of the whole"), the impetus to delay, complicate, put off the flow of energy. All of life participates in, inherits, this initial impetus, which is what all its forms share in common: the impulse to protract time, to age, to organically decay, to exhibit surprise, to function unexpectedly, though perhaps not without pattern or form. This is what life, in its broadest forms, is: hesitation, protraction, waiting, aging, endurance. Adaptation is how particular species and individuals utilize the phenomenon of growth and development, how they use vital force, the excessive energy of the leap needed for innovation and the unexpected, for their own interests. But life as a whole is not adapted any more than it is adapting; it is not compliant with or selected by its environment except negatively (something Darwin himself recognized). In agreement with Nietzsche, Bergson affirms that all that is innovative comes from life itself, through the obstacles it encounters, the material forces it must convert into its own, the inventions, personal and collective, it brings into being. Life is the ongoing tendency to prolong and proliferate change, to delay the expenditure of energy until a chosen moment.

Bergson's Darwin

Creative Evolution was first published shortly after the emergence of neo-Darwinism, which drew together Darwin's account of individual variation and natural selection with Weismann's then recent distinction between the germ plasm and the soma. As discussed earlier, Weismann's hypothesis suggested that the work of individual variation occurs only in the germ, or, as it is now known, the genetic code, produced in the sex cells, spermatozoa and ova, rather than in the biological organism, or the soma. There is an unbroken line from the germ plasm of one generation to that of the next, which is the line of heredity. Implicit in Weismann's position is the belief that organismic development or history, the history of random variations of the soma, has no effect on evolutionary processes, a direct criticism of the Lamarckian doctrine of the heritability of acquired skills. The neo-Darwinist position that developed in Bergson's time, and has grown more powerful since, considers ontogeny to be the combined result of genetic mutation and natural selection, a purely mechanical process that, in principle if not in fact, could be predicted, recognized to have broad directions and knowable constraints, and which could, if it found another bearer or carrier for genetic material and its reproduction, dispense with the biological entity itself. It is precisely this position that Bergson opposes in his account of *Creative Evolution*: the reduction of evolution to the genetic code and the reduction of the organism to its germ plasm function. This position is an early anticipation of Dennett's (1996, 20) reading of Darwinism, now coded in the language of genetics: "The fundamental core of contemporary Darwinism, the theory of DNA-based reproduction and evolution, is now beyond dispute among scientists. It demonstrates its power every day, contributing crucially to the explanation of planet-sized facts of geology and meteorology, through middle-sized facts of ecology and agronomy, down to the latest microscopic facts of genetic engineering. It unifies all of biology and the history of our planet into a single grand story."

Given that neo-Darwinism, since Bergson's time, has overwhelmed the arena of evolutionary biology and threatens to dominate it more and more, it may be now strategically worthwhile to return to some of the more ontologically oriented insights developed in Bergson's rather prescient critique and his insistence on the radical nature of Darwin's own observational hypotheses. Bergson's rereading of Darwin may provide a necessary counterbalance to the dominance of scientific research directed to the various parts or elements of the evolutionary processes by reemphasizing the continuous and indivisible character of biological change. Bergson's argument, in the most simplified form, is that if biological or evolutionary change is broken down into units, parts, elements, or stages, what is fundamentally changing about evolution, its mobility and dynamism, is lost.

Dennett's position identifies the genetic structure as evolutionary cause. This is the basis of Bergson's implicit disagreement with Dennett's algorithmic understanding of Darwinism.[18] As Dennett (1995, 59) understands it, Darwinism implies the functioning of a mindless automatism, a mechanism or algorithm that functions patiently and systematically to sift through all variations, selecting only those of the fittest or the most functional:

"Here, then, is Darwin's dangerous idea: the algorithmic level *is* the level that best accounts for the speed of the antelope, the wing of the eagle, the shape of the orchid, the diversity of species, and all the other occasions for wonder in the world of nature . . . No matter how impressive the products of an algorithm, the underlying process always consists of nothing but a set of individually mindless steps succeeding each other without the help of intelligent supervision; they are 'automatic' by definition: the workings of an automaton." All the processes of species production are reduced to the singular functioning of a generalizable algorithm.

Now, what is significant about the structure of an algorithm is that it is governed by a series of individual steps, procedures that must be gone through one by one. Dennett understands natural selection as this blind and "artificial intelligence" (in contemporary terminology, a Universal Turing machine), an "intelligence" that more or less explores every possible combination to sift through all possible life forms to access the well-designed, the fit, to select them out of all other possible combinations. Bergson's understanding of the continuous nature of evolution seems to pose or perhaps anticipate a series of criticisms of this neo-Darwinist reading of evolution:[19]

1. Dennett mistakes the movement of conceptual possibility with physical existence, which is manifest most readily in his emphasis on the genetic code at the expense of morphological development on the germ line (the "units" of reproduction) rather than on the soma (the "unit" of life). In short, he takes an element of the process of evolution for the process itself. He mistakes the virtual for the possible.[20]

2. Dennett assumes that the relations of variation and natural selection occur step by step, but, as Bergson makes clear, it is only once something is actualized, once it has come into existence, once a first step is completed by a second or a third that we can see it as a step—and even then, it is only a step in reflection, not in actuality. A step, phase, or instant can be isolated only once duration has passed (this is, in essence, Bergson's critique of Zeno's paradoxes, and it can be equally directed to Dennett's position).[21]

3. Even if the step-by-step temporality of the algorithm can be granted, an algorithmic or mechanical process must be able to readily assume fundamental or minimal units. The algorithmic method consists in establishing invariable relations between specific minimal units: it necessarily leaves untouched the degree to which physical existence conforms to these invariable relations (e.g., though we may be able to write an algorithm for playing chess, we cannot determine the moves that any player may make). Indeed,

the distinction between Dennett's position and Bergson's may be illustrated through the game of chess. For Dennett, chess is constituted by all the possible moves of each piece; for Bergson, it consists in the pattern of relations provided by all the pieces. This is why a supercomputer represents for Dennett an ideal chess player, whereas for Bergson, it is the grandmaster, whose intuitive knowledge of the game enables her or him to eliminate most moves and concentrate only on those of strategic relevance. Dennett's intelligence consists in automatism; Bergson's is intuitive. The physical process is open in a way that the algorithmic process is not. This is why Dennett, and the neo-Darwinists whose positions he justifies, are so little interested in the question of development, which both problematizes the notion of a fundamental unit, as Darwin well understood, and precludes an analysis of what is invariable, for there is nothing but endless variation.

Bergson was highly critical of the neo-Darwinians, arguing that, like the operations of classical science represented paradigmatically by Newtonian physics, they tended to reduce evolution to a series of fixed states and the organism to a series of distinct stages, the earliest of which is represented by genetic or chromosomal arrangements: the genetic code. The contemporary thrust of evolutionary theory, since the findings of Watson and Crick on the structure of the double helix and its reliance on the atomism of the individual gene, has been to understand evolutionary processes in terms of the units, now genetic, chromosomal, that compose it. This reduction of processes to states, of life to genes implies that the temporal processes, which are always continuous and not readily explicable in terms of states, are ignored. It is significant that Bergson is highly critical of Weismann's account, which closely anticipates Dawkins's conception of the selfish gene: "Regarded from this point of view, *life is like a current passing from germ to germ through the medium of a developed organism.* It is as if the organism itself were only an excrescence, a bud caused to sprout by the former germ endeavoring to continue itself in the new germ" (CE 27).

Bergson's aim is to steer a course between radical mechanism and radical finalism, to provide an analysis of life in its generality as a mode of strategic, biological overcoming of the material obstacles, a rising to the provocations that the environment, including other species, poses to each individual and species. He claims that we cannot *understand* living systems without either recourse to some purely mechanical base or reference to some external purpose, aim, or goal. We need other terms by which to conceptualize the particular logic of development, of use, and of change that is essential for understanding the characteristics of life. In Bergson's terms, "The hypothesis of purely accidental variation [mechanism] and that of a variation directed in a definite way under the influence of external conditions [finalism]" (CE 62) are equally problematic, for neither has any conception of open-ended becoming.

His own position is developed through a detailed discussion of evolutionary convergence, that is, the development of similar organs in different species, an object of interest and curiosity for Darwin as well. To explain convergent development, there must be some acknowledgment of a (possibly unknowable) common progenitor, whose offspring diverged further and further as time progressed, yet still retained some trace, rudimentary or developed, of their common origin.

Bergson claims that one could never understand the evolution of, for example, the eye in the wide, indeed prolific, range of forms it takes in vastly different species if this were to be explained only in terms of the accidental accretion of random variation. The probability is extremely small that two entirely different species that diverged in their evolutionary development before the emergence of the eye both developed such a remarkably similar and extremely complex and integrated structure and identical function randomly: "The more two lines of evolution diverge, the less probability is there that accidental outer influences or accidental inner variations bring about the construction of the same apparatus upon them, especially if there was no trace of this apparatus at the moment of divergence. But such a similarity of the two products would be natural, on the contrary, on a hypothesis such as ours: even in the latest channel there would be something of the impulsion received at the source" (CE 54).

Change cannot be simply random, as Darwin himself recognized. Most random variation is harmful to the organism rather than productive for it. Only those changes that can be adequately absorbed into the organism as a whole to facilitate or maximize its functioning constitute evolutionary development. Only those changes that constitute a positive production and elaboration of the body count as successful variations. Change is indeed invention and the generation of the new, but the new must remain capable of some mode of assimilation into the already given, or rather, a mode of emergence out of the already given, in coherence and continuity with it, for it to constitute an improvement. For the systematic accretion of a complex arrangement of elements, layers of lens, retina, tissue, optic nerves, neurological connections, and so on, there must be a coordination of changes and their integration with other changes in the body. In other words, the gradual accumulation of random characteristics cannot contribute to the functioning of the organism and the evolution of life unless these characteristics are capable of being both coordinated together and integrated into the functioning of the organism as a whole.

The selective operations of the environment are not able to explain the coordinated changes that make up the evolutionary history of the eye any better than is an endogenous force. The environment, in Darwin's conception, has no positive role to play in the generation of variation, but only in its selection. Natural selection operates negatively, to eliminate organisms or species that are unfit rather than to generate new and successful strategies for survival. In any case, the environment functions at best indirectly: adaptation to an environment does not explain the capacity to generate new and more successful strategies. Neither models of adaptation to an external situation nor models based on the gradual but random internal aggregation of new characteristics can adequately explain convergent evolution or the creation of homologous organs.

Bergson talks about the mutual error of finalism and mechanism: both conflate an organ with a function. Mechanism begins with an organ, the eye, and generates a function from it, whereas finalism begins with a function, seeing, and generates an organ from it. Mechanism must make a staggering hypothesis about the homologous development of the eye in, say, species as widely divergent as vertebrates and the mollusks: that thousands of random variations in divergent paths led to similar structures and functions in surprisingly unlike life forms, accidentally converging on a similar organ through the slow, small, accidental, and gradual accumulation of characteristics. Finalism makes a similarly staggering assumption, of sudden, widespread, and sweeping change: that a few dramatic and coordinated leaps of complexity enabled the organism to be more and better adapted to the environment. In place of both positions, Bergson claims a coordinated simplicity: the eye is neither an accident nor is it merely occasioned by the call of light on the organism. Ironically, although Bergson provides quite strong criticisms of both finalism and mechanism, in addition to the Lamarckian hypothesis of the heritability of acquired characteristics, characteristics acquired by effort, by the soma, and passed on through the germ plasm, he doesn't wish to dismiss these discourses out of hand. Each offers a partial truth, a perspective on an element of the evolutionary process, even though each thereby fails to understand the whole: "Each of them must correspond to a certain aspect of the process of evolution. Perhaps even it is necessary that a theory should restrict itself exclusively to a particular point of view, in order to remain scientific, i.e. to give a precise direction to researches into detail. But the reality of which each of these theories takes a partial view must transcend them all. And this reality is the special object of philosophy, which is not constrained to scientific precision" (CE 84).

In suggesting that the key to understanding evolution, and thus the development of life, ironically lies beyond the scope of biological or scientific analysis, Bergson is not claiming that it is altogether beyond comprehension, but only that new modes of understanding are necessary, in particular, philosophical intuition, which seek access to the whole rather than an analysis of the component parts. He hints at how we might understand the genesis of the eye in a more convincing fashion than prevailing neo-Darwinisms, using the terminology and commonness of both mechanism and finalism. In the act of seeing, we must place simplicity and complexity where they belong: we need to distinguish between the complexity of the organ and the simplicity and unity of its function.

Mechanism addresses the complexity and multiplicity of the parts of the eye: "The mechanism of the eye is, in short, composed of an infinity of mechanisms, all of extreme complexity" (CE 88). Yet it is unable to address how this multiplicity provides any coordination or correlation between these complex parts. Finalism assumes that these parts are correlated through their subordination to a singular, preconceived or externally regulated function. But it cannot explain how this goal is attained except through a certain inevitability or givenness. Neither of these views can ultimately explain evolution because both presume that nature functions like the human being, like an engineer or scientist, bringing together, piece by piece, component parts conceived separately, whereas life functions not through addition and augmentation but through division and bifurcation (CE 89).

Bergson claims that the movement of evolution is extremely simple yet not mechanical. It has a direction and an orientation, but these are not even in principle predictable. Life is always directed to the possibilities of action of whatever kind, whether this is the action of absorption that plants perform or acts of locomotion and muscular exertion in animals. Evolution is the result of the emergence of an impetus of matter to *act* when it becomes cellular in form. This tendency to action, the élan vital, is extremely simple, singular, unified, and indivisible. However, scientific or analytic impulses necessarily resolve these processes—development, growth—into myriad convoluted and complex elements and parts, which, even when put together, never quite yield the unity and cohesiveness of what they attempt to recompose.

To simplify his extraordinarily complex argument, Bergson resorts to an analogy. We need to imagine that the movement of life is analogous to the simple, continuous, and undivided movement of my arm in all the actions it undertakes: "If I raise my hand from A to B, this movement appears to me under two aspects at once. Felt from within, it is a simple, indivisible act. Perceived from without, it is the course of a certain curve, AB. In this curve I can distinguish as many positions as I please and the line itself might be defined as a certain initial coordination of these positions. But the positions, infinite in number, and the order in which they are connected, have sprung automatically from the indivisible act by which my hand has gone from A to B" (*CE* 90–91).

Mechanism considers the various positions through which the arm passes; finalism considers the order of these positions, with the aim of explaining the arm's final resting point. But neither adequately understands the reality that is the movement itself, which is a continuous but ever varying, simple, and indivisible act. The more this simple act is divided, the more stages and the more complexity can be introduced to explain it:

> With greater precision, we may compare the process by which nature constructs an eye to the simple act by which we raise the hand. But we supposed at first that the hand met no resistance. Let us now imagine that, instead of moving in air, the hand has to pass through iron filings which are compressed and offer resistance to it in proportion as it goes forward . . . Now, suppose the hand and the arm are invisible. Lookers-on will seek the reason of the arrangement in the filings themselves and in forces within the mass. Some will account for the position of each filing by the action exerted on it by the neighboring filings. These are mechanists. Others will prefer to think that a plan of the whole has presided over the detail of these elementary actions, they are the finalists. But the truth is that there has been one indivisible act, that of the hand passing through the filings. (*CE* 94)

In the process of dividing this simple movement the act is thereby transformed, and the most real element—its dynamic of change, its durational protraction—disappears. Time is resolved into space, movement is reduced to points at rest. So it is with the path that life takes in its global movement, which Darwin called evolution: life is a process that is neither directed from some internal purposiveness or functionality (whether this is understood as radical finalism, as humanist teleology, or whether it is assumed in the form of genetic reproduction, or the genetic programming of characteristics), nor is it simply a reaction to external impingements. Rather, it is a singular process that diverges from itself in its growth and development. So, too, with the evolutionary development of the eye:

> According as the undivided act constituting vision advances more or less, the materiality of the organ is made of a more or less considerable number of mutually coordinated elements, but the order is necessarily complete and perfect. It could not be partial because . . . the real process which gives rise to it has no parts. That is what neither mechanism nor finalism takes into account, and it is what we also fail to consider when we wonder at the marvelous structure of an instrument such as the eye. At the bottom of our wondering is always this idea, that it would have been possible for a *part only* of this coordination to have been realized, that the complete realization is a kind of special favor . . . No matter how distant two animal species may be from each other, if the progress toward vision is gone equally far in both, there is the same visual organ in each case, for the form of the organ only expresses the degree in which the exercise of the function has been obtained. (CE 95–96)

In other words, each organ always and only functions to help engender and make more accessible the world of objects, the world in which the organism lives and in which it must maintain itself. Light no doubt provides a kind of provocation to cells to somehow register and harness it; it acts as an external stimulus, a resource capable of utilization only if certain cellular changes and biological arrangements occur. But equally, cells, which have an internal tendency to unite with other cells in particular arrangements to form organs and enable biological syntheses, are modes of active exploration and use of the world, including the photosynthesizing and the vision-enabling properties by which light comes to us. During development, cells are organized to produce specific organs, and organs are organized to produce specific functions, and each organ performs its role in the collective integration of the body (to this degree, Bergson affirms both mechanism and finalism). They each have a direction, but that direction is given only within the whole and is never simply a component of the whole.

Life, in this sense, is the response to the matter that gives it life and sustains it as such. Life is not an adaptation to the external environment, a giving in to or conceding of the primacy of the environment, a passivity in the face of its activity. Rather, for Bergson, life is a response, a reply, to the problem of how to live in a material world and to use its resources to maximize itself, to maximize its facilities and capacities, to go as far as it can in the directions virtual to it:

> Life is, more than anything else, a tendency to act on inert matter. The direction of this action is not predetermined; hence the unforeseeable variety of forms which life, in evolving, sows along its path. But this action always presents, to some extent, the character of contingency; it implies at least a rudiment of choice. Now a choice implies an anticipatory idea of several possible actions. Possibilities of action must therefore be marked out for the living being before the action itself. Visual perception is nothing else: the visible outlines of bodies are the design of our eventual action on them. Vision will be found, therefore, in different degrees in the most diverse animals, and it will appear in the same complexity of structure wherever it has reached the same degree of intensity. (CE 96)

The eye emerges as an organ of vision in a number of apparently unrelated species, not because they share common raw materials or origins, though all life does share the unity of a common origin in self-differentiation that diverges more and more as time passes (CE 103), nor because of the common influence of the environment, in which visible light may well function as a primary vector or organizing force in many unrelated species, but because the structure of the eye in all its multitude of variations (the eyes of an insect do not resemble those of vertebrates, for example) is an active response to a problem, the problem of how to act amid matter. It is not the response of a conscious subject, an individual, but of a species, a morphology, a bodily form. It is a remarkable and inventive response to the problem that the virtuality of movement poses to living, moving beings. At every stage in the evolutionary development of the eye in a particular species, it is adapted, it functions to facilitate survival: there is no partially working eye. The pigmentary masses in lower organisms function for those organisms as well as does the eagle's eye. Whereas organs are clearly of unequal complexity in different species, each visual apparatus is equally coordinated with and integrated into the other bodily activities of each species: "The form of the organ only expresses the degree to which the exercise of that function is obtained" (CE 96). The more locomotion in and action on the material world is facilitated, the more successful are the survival strategies of the living being, the better adapted it becomes. Vision is one of the more successful and varied strategies life has developed to facilitate action; it provides one form of outline of the possible action of objects. The eye broadens, widens our sphere of action, but more, it provides time, delay, and thus choice in the possibilities the visible object affords for future action.[22]

Some Implications

We will return to the divergent lines of torpor, instinct, and intelligence along which life elaborates itself in the next chapter, when we link them to the question of intuition. But for the remainder of this discussion, it is necessary to draw some connections between *Matter and Memory* and *Creative Evolution*:

1. The universe as a whole has duration and is subjected to an "inner dynamic" of constant change, though material systems which we can cut out of its continuity may be rendered closed systems and thus subjected to very precise measurement and predictability. This duration, however, makes itself actual, instead of subsisting virtually, only with the exercise of life itself, which embodies and elaborates the forward pull of duration. Life is the most elementary expression of duration, the prevalence of the future over the past and present, which is latent or submerged in the nonorganic universe.

2. Evolution is a surprise, an invention, though it is not the invention of any thing or any one. Life is what centers the universe around the needs and concerns of living bodies, each of which outlines the world and its objects according to its interests, extracting from those outlines what it needs to survive and develop. Life introduces indeterminacy into the universe in this outlining or highlighting of material objects, which is also an obscuring of their full and determinable effects, the capacity to place them in unexpected contexts and extract from them unexpected uses, their capacity to make new use of objects and new objects from given materials in the future.

3. Life proceeds or develops in duration through differentiation and division. An initial simplicity and unity of functions and modes of connection diverge over time, producing increasingly different lines of development, increasingly complementary and antagonistic species, which grow further and further apart in morphology and functions as time progresses, which develop increasingly specialized functions not only for their environmental niche but for indeterminable transformation in directions latent in but not visible to present forms. As life forms become more specialized, their difference becomes more distinctive.

4. Life is constituted, not of differences of degree, one creature being a version or development of another, but of differences in kind, the result of

natural articulations. Although life is itself a difference in kind from matter, the elevation of matter above its expected expenditure of energy, it is itself divided into mutually exclusive trajectories that nonetheless share a common source and thus lie dormant within each other, vegetative or animal, centralized or decentralized nervous system, instinctive or intelligent action, the lines of difference in kind in life itself.

5. Life is the simultaneity of a virtual and an actual, past and present, matter and memory. All life evolves from earlier life and contains within it certain residues or living memorials (e.g., mitochondrial DNA) of its biological past. Not only are present forms of life the actualization of some but not all of the virtual tendencies latent in their progenitors; they also contain within themselves a virtuality that varies itself in its process of actualization, that is, in the continuing movement of evolution that ensures our progeny will differ too in ways not explained by but already to some extent contained within our present forms.

6. The evolution of life is larger than and contains within itself the knowledge or intellectual apprehension of life. Intelligence is not outside the evolutionary process but is its historical product, which is why the traditional techniques of science are both so well suited to the apprehension and manipulation of matter but so unsuited to understanding the movement of life. They, too, are evolutionary products, the attempt by specific creatures to carve up the real in terms of its interests.

In the next chapter we will explore in further detail the elaboration of animal life in the increasing divergence of instinct and intelligence, link it to the relations between epistemology or forms of knowing and ontology or forms of becoming, and to intuition as the philosophical method by which what has been divided—instinct and intelligence, knowing and being—can be reconsidered in their intimate connection.

9. INTUITION AND THE VIRTUAL

Nature has no evolutionary plan, purpose, goal, or function. This is the basis of Bergson's critique of finalism. It is only in retrospect, after we provisionally pause to look backward, that is, create a concept of history, that any clear direction or path can be discerned. Life is not conformity to some natural plan, for any plan is, by definition, the attempt to foreclose certain options, chances for the virtual to actualize itself, the cutting off of future openings, a technique of elimination, the giving in advance of a future that has yet to be made. A plan implies the elimination of duration, the compression of the future into the present. Life itself, however, functions not through conformity to a plan, an ideal, or a law, but through processes of differentiation, whose "plan" or direction is only emergent, in the process of being developed.

The Divisions of Life

For Bergson, life itself is an utter contingency: it need never have emerged, and there is no particular explanation for why it developed in the forms it presently takes on earth. He remains open to the possibilities of other life forms, other modes of emergence, other techniques for life's eruption into the universe of matter than those with which we are familiar. Indeed, he actively conjectures about how it is possible to conceive life outside the carbon-based model that marks terrestrial existence. Like contemporary theorists of artificial life, Bergson is fascinated by the historicity, the accidental characteristics and material forms of life on earth, though unlike them, he is not interested in undertaking the simulation of life in computer programs or scientific experiments, though no doubt he would be interested in the results of these contemporary experiments.[1] As he suggests:

> Two things only are necessary [for the emergence of life]: (1) a gradual accumulation of energy; (2) an elastic canalization of this energy in variable and indeterminable directions, at the end of which are free acts.
>
> This twofold result has been obtained in a particular way on our planet. But it might have been obtained by entirely different means. It was not necessary that life should fix its choice mainly upon the carbon of carbonic acid. What was essential for it was to store solar energy; but, instead of asking the sun to separate, for instance, atoms of oxygen and carbon, it might (theoretically at least, and apart from practical difficulties possibly insurmountable) have put forth other chemical elements, which would then have had to be associated or dissociated by entirely different physical means. (*CE* 255)

There has been tremendous interest in recent years in the simulation of life forms through the creation of computer programs that exhibit some of the characteristics of learning, development, surprise, or evolution. Increasingly elaborate programs that focus on reproducing or generating swarming behavior, the emergent patterns of movement and behavior that large numbers of undirected individual units produce without a central director, a set of external rules, or conscious coordination, have come closer and closer to generating recognizable patterns of animal behavior. Craig Reynold's "boids," computer-generated flocks, herds, and schools of swarming units (whether of birds, ants, bees, sheep, or fish, for example), emulate amazingly life-like patterns through the production of very simple algorithms, or basic principles.[2] The study of emergent behavior, qualities, and effects that develop not through planning and coordination but through bottom-up, local actions shows that passing a crucial threshold, a phase transition in which effects amplify themselves, produces unpredictable outcomes of surprising complexity, whether this kind of study is conducted sociologically and in urban studies as the analysis of the phase transition between the town and the city, or biologically, in the study of how cells develop embryologically into coordinated organs and bodies, or in other fields.[3] While the simulation of patterns of emergence is becoming more and more sophisticated, it should be remembered that what is produced are representational patterns, not living forms, though they resemble the patterns that living forms take: the simulation remains a semblance or image, a second-order patterning that derives from living forms some of their living characteristics but also transforms and rigidifies into automatic processes, algorithmic steps, that which, in the natural world, has no steps, no gaps, no halts. Other, new forms of life, silicon life, may indeed have been created on computer screens over the past few decades, but it remains to be seen to what extent these artificial entities, which are nevertheless still in some sense planned (i.e., programmed), exhibit evolutionary autonomy, that is, to what extent they resemble the less visible characteristics of the living.

Life as we know it is the production of divergent pathways that one could understand as a primitive mode of choice. There are alternative directions and orientations that life, at its various points of transformation, at key moments of phase transition, faces. These are not conscious choices, of course, but ways of living. These alternatives, as Bergson argues in *Time and Free Will*, are not pregiven options; they are virtualities, tendencies, which all of life shares but which only some species actualize, each in its own way. The most significant bifurcation in evolutionary development, Bergson suggests, is that between what he terms "torpor" and locomotion, which constitutes the most basic rift between plants and animals. At a certain evolutionary moment, when the earliest forms of life emerged and developed, the natural world afforded many possibilities of sustaining existence, but life itself devised only two. The sedentary form locates the living being in a fixed environment and requires it to sustain and reproduce itself using techniques to extract and utilize the materials given to it in its location (carbon and nitrogen, in particular). The vegetative world, which first combined the extraction of carbon and nitrogen in single organisms, divided and became more and more specialized, dividing itself (again) into a cellular structure specialized in the absorption of either carbon or nitrogen. Plants retain the capacity, through the synthesis of carbon, to store energy; "microscopic vegetables" (CE 117; what today we would call microbes or bacteria) convert nitrogen from the air, soil, and living beings. Bacteria divide themselves into different, sometimes competing species through an increasing focus on particular modes or objects of extraction. The vegetable kingdom, one of the primary lines of bifurcation of life, itself becomes divided into increasingly dissociated, differentiated species, growing further and further apart as time progresses, dividing itself along a carbon line and a nitrogen line, a line of torpor and a becoming-animal that never attains the form or structure of the animal.

The animal form, though it cannot extract the resources it needs directly from its mineral, liquid, and solar location, makes use of plants' capacity to manufacture nutrients from their given surroundings, which in many cases requires microbial intervention to help facilitate processes of ingestion and metabolism. Mobile life must direct itself to acquiring less secure, less guaranteed, but also more varied sources of nutrition, and with it, the capacity to both discern the value of this variety of resources and to access those resources by a more and more enhanced mobility. Self-regulated mobility implies the dawning possibility of choice: "If consciousness means choice and if its role be to decide, it is unlikely that we shall meet it in organisms which do not move spontaneously, and which have no decision to take. Strictly speaking, there is no living being which appears completely incapable of spontaneous movement. Even in the vegetable world, where the organism is generally fixed to the soil, the faculty of movement is dormant rather than absent: it awakens when it can be of use . . . It appears to me therefore extremely likely that consciousness, originally immanent in all that lives, is dormant where there is no longer spontaneous movement" (*ME* 10–11).

If torpor or unconsciousness and movement or consciousness (however rudimentary: consciousness is as complex as the avenues of "choice" available to the living being), that is, becoming-plant or becoming-animal, are the two lines along which life (contingently) develops, these two tendencies, Bergson suggests, remain virtual in all living beings and testify to their common ancestry. There is still a plant-becoming in all animals; it is only through effort and activity, "at the price of fatigue" (*CE* 113), that the animal avoids the danger of unconsciousness and eventual death that marks its own vegetative existence. And there is, perhaps, an equal, if equally latent or rudimentary, impulse to incipient movement in vegetative life, as witnessed by the evolutionary development of carnivorous plants and the emergence of vines and creepers, plants with a limited capacity for movement.[4] The sleep-like immobility of plants has become latent in most animals (Bergson makes echinoderms and mollusks, animals with limited control of movement, contemporary exceptions), and the striving to explore and expand movement is latent in most plants, yet taken together, they testify to a common ancestor in whom both of these tendencies were virtual. This makes torpor and consciousness both complementary and mutually exclusive: they are "options" that come from the same source yet are divergent paths where participation in one more or less excludes the other.[5]

Plants acquire a cellulose structure which then enables their roots to extract minerals and their foliage to extract and store radiant energy through the chlorophyllic function of photosynthesis, and to expend this stored energy in the slow, continuous processes of growth, regeneration, and reproduction. Animals, even with elementary nervous systems, by contrast, require the stored-up potential energy of plants (and sometimes other animals) to produce the particular distribution of energy that is less even and measured and more explosive, more able to generate rapid and unpredictable movements. Animals require what Bergson refers to as an "explosive energy" to execute movements, to exercise their options. This explosive force, which Bergson likens to the unstable functioning of gunpowder (*ME* 14), is derived from animals' capacity to store, at least temporarily, the energy contained in carbohydrates and fats, that is, energy derived from plants and other animals. This is, indirectly, the energy of the sun and of minerals, processed by plants before they can be absorbed by animals, the carbon cycle of life on earth:

> To execute a movement, the imprisoned energy is liberated. All that is required is, as it were, to press a button, touch a hair-trigger, apply a spark: the explosion occurs, and the movement in the chosen direction is accomplished. The first living beings appear to have hesitated between the vegetable and animal life: this means that life, at the outset, undertook to perform the twofold duty, both to fabricate the explosive and to utilize it in movements. As vegetables and animals became differentiated, life split off in two kingdoms, thus separating from one another the two functions primitively united. The one became more preoccupied with the fabrication of explosives, the other, with their explosion. But life as a whole, whether we envisage it at the start or at the end of its evolution, is a double labour of slow accumulation and sudden discharge. (*ME* 14)[6]

The energy of the sun becomes stored as potential energy in living forms, through the photosynthesizing activity of plants. When animals ingest plants, this initially solar energy is converted into kinetic energy through the explosive impact that the animal utilizes to move and to act. Matter continually unmakes itself insofar as the principle of entropy regulates it; life is a struggle to provisionally remake matter, through prolonging the present into an unforeseeable future, endowing matter with a virtuality only life can bring it. Life becomes the way that matter, as energy, prolongs itself: living beings become the way that the energy of the universe is redistributed outside its predictable limits. It is the introduction of a spark of indeterminacy into an otherwise grindingly determinable universe. It introduces a retardation in the flow of energy, a movement mimicked in the cerebral interval of the human brain, a delay that enables alternatives not available to the inorganic world to emerge for the living being that inhabits and makes that inorganic world a virtual reservoir of its future actions (*CE* 246).

Before life is divided into vegetative and animal streams, Bergson hypothesizes, its most elementary function is the combined plant-like storage of energy with the goal of its animal-like explosive discharge. It is as if matter posed two alternatives to life-to-be: the gradual accumulation of energy and its slow continuous discharge in less effective and less dangerous action (regeneration), or its rapid discharge in more efficacious and more risky action, which has the virtue of a wide range of possible objects and interests not contained in an immediate environment (movement). Life solved the dilemma of these alternative energy-storing and -expending options by irreversibly dividing itself into complementary yet contrary tendencies: sedentary and mobile forms.

The division of life into vegetative and animal series is perhaps the most monumental of the "decisions" or strategies that life invents for maximizing itself, for proliferating and organizing itself into greater and more complex forms of harnessing of energy and materiality. The evolution of the animal world is an intensification of the march of mobility. Glancing over the paleological records, Bergson suggests that the evolution of animal species from the time of their division from floral species can be seen in retrospect as the invention of more complete and novel forms of mobility.

The most primitive forms of animal, from the amoeba to prehistoric worms, Bergson suggests, had a simple shape or form open to the remarkable variety of future forms, indeed an "infinitely plastic" (*CE* 130) shape and potential, insofar as these creatures were the earliest ancestors of all subsequent forms of animal life. The earliest animals, he claims, needed protection from their environment, and the more protected, the less facilitated was their mobility. The earliest mollusks and arthropods were clad in protective coverings—the shell for the mollusk, the carapace for crustaceans, a bony sheath for many fish—which facilitated a greater rate of survival against environmental dangers and catastrophes but also proved in the long run to be safer yet less successful in inventing new forms of existence, in maximizing their capacity to mobilize themselves. The protective covering, Bergson conjectures, may have been the evolutionary consequence of the emergence of carnivorous animals, those whose enhanced mobility enabled them to prey on other animals as well as plants, and which, through natural selection, preserved and privileged those that developed bodily defenses against such predation. While protecting these early animal species with a natural covering, their diminishing mobility cast them toward the tendency to torpor or unconsciousness: "The hard and calcareous skin of the echinoderm, the shell of the mollusc, the carapace of the crustacean and the ganoid breast-plate of the ancient fishes probably all originated in a common effort of the animal species to protect themselves against hostile species . . . If the vegetable renounced consciousness in wrapping itself in a cellulose membrane, the animal that shut itself up in a citadel or in armour condemned itself to a partial slumber. In this torpor the echinoderms and even the molluscs live today" (131).

When fish evolution replaced the bony breastplate with scales, or insects their exoskeleton with articulated limbs, their increasing vulnerability to predation was compensated through their increasing agility and mobility. An evolutionary risk occurred: by freeing creatures from their bodily armor, they risked extinction; but by mobilizing themselves in more and more adept forms, they counteracted this risk with the rapid development of the skills of suppleness and movement. Movement and suppleness themselves help specify the growing division within animal species, the dawning bifurcation between arthropods (or insects) and vertebrates, that is, between creatures in whom the body is formed through the addition of limbs that distribute and organize movement among many appendages through a decentralized nervous system, and creatures for whom the body resolves itself into two sets of limbs that require the coordination of a central nervous system.

The insect world becomes governed by the operations of increasingly complex forms of instinct. The world of vertebrates, by contrast, becomes more and more marked by the functioning of an incipient intelligence. Thus, if vegetative life can be characterized by torpor and unconsciousness, animal species can be described in terms of a widening rupture or division between instinct and intelligence. Torpor, instinct, and intelligence are the three lines of emergence and development that mark life on earth, the three vectors that orient life's material forms. Bergson is not suggesting a linear movement through three "stages" of evolutionary development: torpor does not give way to the instinctive activity which then develops itself into intelligent activity. Rather, life divides itself into possibly complementary and usually potentially antagonistic or rivalrous relations between these three tendencies (see figure 3).

Life itself is essentially fanning out, differentiation, division.[7] Significantly, these temporal features characterize not only all living beings; they can also be attributed to material systems, natural systems outside the experimental control of variables under scientific conditions (the weather,

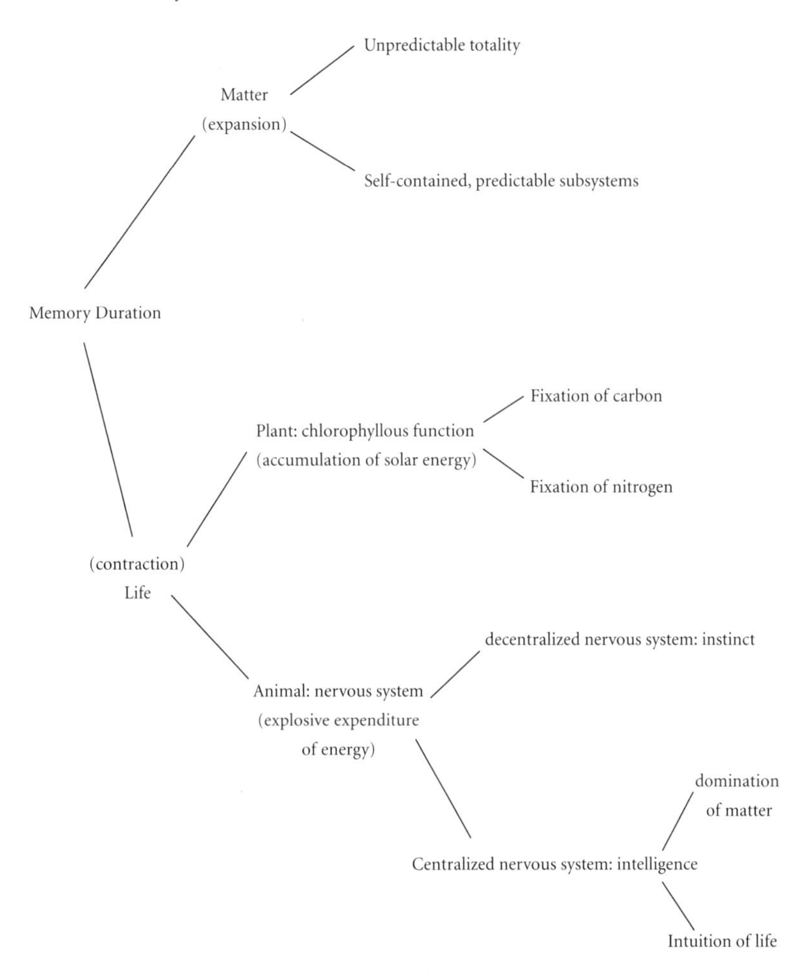

Figure 3. Diagram of Differentiation
Source: This is a modification of Gilles Deleuze's diagram in *Bergsonism*, trans. Hugh Tomlinson and Barbara Habberjam (New York: Zone, 1988), 102.

large-scale geological shifts, the spread of epidemic infections, etc.). The interpenetration of states, the arbitrary nature of divisions and cuts in natural systems, the fundamentally unpredictable character of change, the fact that change occurs as a kind of internal growth rather than an external upheaval, are as relevant to our understanding of nonorganic natural systems as they are of organic ones.

The tripartite division between torpor, instinct, and intelligence, by which Bergson identifies the divergent evolutionary lines of vegetal, arthropod, and vertebrate existence, is a distinction between three different types of relation of the living being to its material world. The vegetal line describes a relation of automatism: the plant, rooted in its environment, automatically extracts what it needs from the soil, the air, light, and water; lacking mobility, it has a minimal or dormant consciousness. Unable to undertake voluntary movement, the plant makes no "decisions," undertakes no mobile actions: it is unconscious. The insect line is directed to objects and the world through a different and much more complex automatism, now defined not by its fixed location but by a given repertoire of responses to triggering situations. Insects function through bodily impulses that exercise both a degree of freedom and a degree of constraint. They thus have consciousness, though its requirements are rudimentary, linked largely to the recognition of instinctive triggers. The vertebrate line, functioning through the primacy of intelligence, generates new activities and innovative patterns for the utilization of objects. It is the line in which consciousness is the most elaborated, the brain and central nervous system being the organs that most facilitate delay, and where freedom of activities is the most pronounced. The next section examines in more detail the functioning of the insect-instinct and vertebrate-intelligence relation.

Instincts

Because instinct and intelligence share a common origin, their separation is never final or complete. Even where there is a clear preponderance of one, the dormant possibilities, the virtual existence of the other lies alongside it. Just as mobile and sentient beings contain within themselves the submerged tendency to torpor and unconsciousness, so instincts also contain the rudiments of intelligence, and the functioning of intelligence has instinctive components. The differences between plant and animal, and between insect and vertebrate, are not differences of degree but differences in kind because, although they share a common origin, they use their inherited resources in fundamentally different ways. These are among the "natural articulations" of life: "The group must not be defined by the possession of certain characteristics but by its tendency to emphasize them" (CE 106). There is no form of intelligence that does not bear some traces of rudimentary instinctive acts, and there is no instinct without what Bergson calls "a fringe of intelligence" (136). Although they are different mixtures, instinct and intelligence are not the development or arrest of each other: instinct is not atrophied intelligence, and intelligence is not a conscious form of instinct. They diverge from each other more and more over the passage of time, developing different aims and functions. They become, in use, mutually exclusive and potentially antagonistic. They are not specific attributes or qualities of various species, but tendencies, directions, trajectories, emphases, or orientations.

Instinct is at its most developed, Bergson claims, in the arthropods, and the most complex and developed of the arthropods are the social insects: bees, ants, wasps, and so on. Their highly complex repertoires of behavior involve a considerable range of choices and multiple possibilities for action, even though we consider that they are all largely triggered by instincts, by behavior generated through given cues. Their behavior is not learned, but it does involve a direct awareness, an immediate knowledge of living things and those elements in the environment that each requires for its sustenance and reproduction. Bergson illustrates with the well-known examples of the anatomical "insights" of the wasp, which somehow, instinctively, knows just where to paralyze its prey so that it is not killed but incapacitated, and thus provides the wasp with living, ongoing sustenance.[8]

Instinctive activities are not simply automatic responses to a pregiven trigger, for they involve the discernment and attunement to the particularities, and not just the general form, of the stimulus. Take the case of bees: "When by exception they build in the open air, invent new and really intelligent arrangements to adapt themselves to such new conditions," and even in situations where their usual hive-building activities function unimpeded, bees exercise intelligence in the "choice of place, time and materials of construction" (CE 142). There is, in short, a small measure of "freedom," of the uninstinctive, or of the intrusion of a kind of intelligence into the instinctive.

What exactly is this instinct that directs bees, ants, and wasps with such intimacy and refinement to the contours of their environment and to the detailed physiology of their prey? Bergson claims that we cannot understand instinct in terms of habit, that is, a learned chain of responses to a given situation. Habit is a modality of intelligence. Instinct, for him, is closely linked, not to a compressed and synthesized chain of (past) activities reanimated in the present, but to the development and specialization of bodily movements, to the elaboration of the insect's bodily morphology or morphological potential: "The most marvelous instincts of the insect do nothing but develop its special structure into movements" (CE 140). Instincts are the mobilization of organs and bodily attributes into focused and concerted action. Where there is a social division of labor through different "classes" of insect, there is a corresponding morphological difference: different instincts are distributed in different bodily forms. This is really another way of saying that insects use their bodily forms as tools. They have enough "intelligence" to use organized tools, and the tools they have at their disposal are ready-made and unalterable, in the form of their own bodily organs and attributes. The intelligence of insects lies in their capacity to utilize the properties of their own bodies to attain their needs from the environment: "Instinct finds the appropriate instrument at hand: this instrument, which makes and repairs itself, which presents, like all works of nature, an infinite complexity of detail combined with a marvelous simplicity of function, does at once, when required, what it is called upon to do, without difficulty and with a perfection that is wonderful" (140).

Perfect in their capacity to direct a species to their privileged objects through the exercise of a particular bodily organ or structure, instincts remain invariable in their operations. In instincts, nature manufactures for the living creature a dynamic and functional instrument, the apparatuses of its body. If the bodily instrument is modified, so too is the instinct and, in the long run, the species. Instinct is the utilization of a specific instrument, a part of the body, for a specific task and a concrete object. Bodily organs are the most adapted to and honed by their environment; form and function correlate perfectly. They represent an ongoing successful strategy for the attainment of specific ends.

Consciousness emerges for the insect, not in the perfect functioning of instinct, but in its thwarted operations. It is when the instinct fails to attain its end that consciousness appears. Consciousness, then, is the result of the deficit of instinct; its appearance is "accidental," epiphenomenal. Consciousness focuses only on the initiation of the instinctive activity, the trigger that releases a series of automatic movements, rather than on the attainment of an end. Instinct is not outside knowledge. However, the knowledge involved is acted rather than represented, and thus largely unconscious. It reveals itself as knowledge only in its failure. In its usual operations, instinct is the utilization of bodily organs in a given manner for the attainment of specific ends, the accomplishment of certain acts. (This, incidentally, is also true of habitual behavior: consciousness appears only if its means or ends are thwarted.) The obstacle that prevents the enactment of an instinctive mode of behavior is what generates the necessity of some intervention, some consciousness, some reorientation of activity.

The knowledge, if it can be called that, that the insect has of its objects of interest is innate, directed to a specific object (or category of objects) of immediate relevance to the exercise of its activities. Implicit rather than represented, the trigger of action rather than reflection, this knowledge is not capable of generalization from one category of objects to another. The same actions performed on the "wrong" objects will produce a failure of ends rather than a modification of means. The moth trapped behind a pane of glass will bang against it over and over in attempting to follow light. Instinct does not learn. It does not carry its past along with its present in order to modify its future, but compresses the past into activities in the present in order to be able to repeat them when needed.

Instincts are no more evolutionarily stable, no less subject to the movement of duration and self-overcoming than intelligence. This explains both the remarkable variety of instinctually regulated behavior in different species, and also the durational or forward direction of this evolutionary movement. There are, of course, many insects whose form has remained relatively unchanged for millions of years; nevertheless, most insects are evolutionarily volatile and subject to unexpected or unanticipatable transformation. Instinct itself seeks its maximum effective variations; it becomes other than its past.

To summarize what, for Bergson, are the key characteristics of instinctive existence:

1. Instincts are a chain of preset behavioral activities triggered by the recognition of certain relevant characteristics in the insect's environment.
2. They do not function altogether outside of intelligent behavior; rather, in making action dominant, intelligence is rendered dormant. Instinct acts its knowledge rather than represents it. Intelligence is induced, though not actualized, both in the beginning of or in triggering instinctive behavior, and by its breakdown. Instincts are fringed by intelligence, but intelligence functions only as an augmentation of or supplement to instincts.
3. Instinctive behavior is based on the exercise of specific bodily functions or activities for the attainment of specific pregiven objects and goals. An alteration in bodily morphology would necessarily propel a different evolutionary elaboration of instinct, a different set of activities, a different exercise of organs.
4. Instinctive behavior is not opposed to the use of tools or instruments. But the instruments it utilizes are not separate from the body itself. Thus, its instruments are perfectly attuned to their objects, having been evolutionarily developed in conjunction with them. But for this reason, they tolerate little variation and lack the flexibility, and virtuality, of the less organized and more constructed tools that interest intelligence.

5. Instinctive behavior is based on a specific relation between the insect-subject and its given range of objects. Its knowledge is thus highly particular, a knowledge of a specific (category of) object that interests it.

Intelligence

When Bergson discusses the evolution of intelligence, what he has in mind is not a particular privileging of human forms of intelligence, reason as the pinnacle of intellectual development (though for him this is true), but rather a practical or active facility for choice. The reflective capacities of reason and introspection are later evolutionary developments that come after the division of arthropods and vertebrates. By intelligence he means the ability to choose among alternative behaviors, the capacity to discern a preference for one mode of activity over another in a particular situation, the capacity to reframe a situation through the use of external tools. It is an intelligence shared by, among others, dogs, cats, monkeys, dolphins, and humans, the point of connection between the human and the animal rather than, as in Aristotle, the point of their division. Nevertheless, as instinct is developed to its highest known degree in the social insects, so too intelligence finds its clearest expression in human invention.

Instinct and intelligence, sharing a common origin, necessarily share common ingredients, even or especially as they are qualitatively different mixtures. Like instinct, intelligence is an attempt to extract things of interest for the living being from the objects of the world. They are both, like automatism, "inventions" of life, the ways that different forms of life make themselves in extracting something they require from the world; they are equally measures of life's diverse strategies to overcome itself, to elaborate and make itself more. Like instinct, intelligence is a tendency rather than an accomplishment, a movement rather than a property or faculty; like instinct, intelligence also works with tools to facilitate life's access to objects. But unlike instinct, intelligence either finds its objects outside the body, in the things in the world, seeing in those objects other uses than those provided by nature, or it produces, manufactures, the tools it needs. Like instinct, there is something innate in intelligence, but what is innate, for Bergson, is a knowledge of relations rather than the instinctive knowledge of objects and their connections with the body.

Instinct shares with intelligence a kind of knowing, yet instinct is a noncognitive awareness of life. Intelligence is the capacity to shape and remake nonliving matter. If instinct insinuates itself into the details of life, intelligence directs itself outward to the regulation and ordering of the material world. This means that, in a sense, the noncognitive functioning of instincts is more at home in life than in matter; ironically, intelligence is more comfortable in matter than in the contemplation of life.

Like instinct, intelligence is a way of dealing with the exigencies of life, a strategy of survival that has been historically successful. Intelligence functions, in the first instance, Bergson asserts, to vary instincts (*CE* 138) and thus to prepare for the creation of artificial instruments through which it can extend its bodily access to the world. Bergson affirms the fundamentally prosthetic nature of intelligence, its close and continuing tie with material invention, with the unexpected use of what is found and its transformation into what is useful. Intelligence and invention are aligned. Invention in fact accelerates the rate of change beyond its biological speed, for tools, invented and used for specific purposes, are superseded or incorporated at a much greater speed than biological evolution.[9] This change in material environment is precisely the invention and intervention of the customary, the social, and the historical into the biological, something for which the biological prepares itself through the elaboration and development of intelligence. Bergson sees in intelligence not only the capacity to manufacture artificial or constructed objects but, more significant, the ability to manufacture or invent tools to make tools, that is, the capacity to infinitely vary what constitutes a tool and what part of the material world, removed from its natural context, can become of direct use in mediating between the body and the world. The body, not prepared or organized in advance for its relevant conjunctions with matter, is now amenable to conjunctions never considered before.

The insect finds its tools readily given in its bodily appendages, and instinct is the lived knowledge of how to use these tools with particular objects in concrete circumstances. In the case of intelligence, which manufactures or finds tools, the tool's capacities and usefulness must be learned. Unlike bodily organs, that is, the tools of bodily morphology, the manufactured tool is always imperfect; it requires training the body to modify its activities to accommodate the tool's potential uses. The body must transform itself, at least temporarily, through the prolonged use of an otherwise extraneous device. The manufactured tool is a reconfiguration of (relatively) unorganized matter, whereas the insect's bodily extensions are always modeled on the specific characteristics of its bodily arrangement; it has an immediacy of effect and is always correlated with its task (see Leroi-Gourhan 1993). The manufactured tool is open to any form, any possible use, any project; unlike instinct, it extends and modifies the body rather than conforms to it; also unlike instinct, it not only satisfies our needs but creates new ones.[10]

Vertebrates open themselves up to a path of evolutionary development that now includes not only the natural environment, but also a social, cultural, and technological universe, one open to continual augmentation and transformation by conventions and inventions. This spreads the scope of their possible interest and action more widely than a given habitat. The environment is now extended in its scope and range and open to concerted change, to new possibilities they invent. Instincts may function to cohere the social organization of insects, but the products of instinctive activity do not accumulate a deeper or more varied knowledge: they produce a deeper and more complex object. What is significant about the inventive fabrication of intelligence is that it consolidates itself socially, through the intelligent functioning of other individuals: it is cumulative and collective, it operates through networks and connections. No honey bee could survive without the collective efforts of others, and the hive is constructed as if the bees functioned as a single, collective organism, coordinated and hierarchically unified by a shared purpose. Although intelligent beings function socially and collectively, there is never this melding of goal and orientation; there is confrontation, disagreement, transformation, compromise, and all the other variations of collective organization, but never this singularity of purpose and activity. On the other hand, the coordinated cooperation of units, Bergson suggests, appears at the level of the cell itself, and it is the cooperative and near perfect functioning of cells that makes the organism possible.

The biological units that make up the body—cells, tissues, muscles, organs, and so on—function instinctively, even in intelligent beings. Cells act as if they were instinctual organisms; they are insect-like in their collective relations to tissue building. They have a "knowledge" of what they are supposed to do, how they are supposed to function, without learning. They "know" in the sense of being able to enact precisely the biological functions they are required to: "All goes on as if the cell knew, of other cells, what concerns itself; as if the animal knew, of the other animals, what it can utilize—all else remains in the shade" (CE 167). Cells function through "instinctive" procedures that enact a knowledge without representing it, just as other instinctive activities do. This is their "sociality," the mutual cooperation of knowledge of the relevant effects on other units (whether cells or individuals) with each one's activities, a precoordinated cohesion under a single model. The sociality of intelligent beings, however, is linked to the possibility of the proliferation of new and surprising models, a multiplicity of ends and goals, which cannot be agreed upon, which cannot be known in advance. Nevertheless, Bergson claims, this instinctive "knowledge" is the consequence of a fundamental unity (perhaps even a biological memory, as some theorists suggest)[11] of life in its evolutionary movement, a fundamental sympathy of all beings with the populations with which they engage.[12]

Consciousness and intelligence open up the material world to the play of virtuality. Consciousness highlights our possible action on things; it is a measure of our virtual interest in things, the gap between the thing and its (newly emergent) uses. It measures the difference between the real object and the potential to use it in a variety of unexpected ways. Consciousness, and intelligence, which is its correlate, are bound up with representation, which delays, complicates, and frees behavior, even as it inhibits it (the more automatic the behavior is, the less consciousness occurs or is required). This subtracts from the object most of its features and details, but it also adds to the object the possibility of new connections, new contexts, new uses.

Where instinct provides an innate knowledge of those objects that are bound up with or relevant to the evolution of insects, there must also be something innate, virtual, about the operation of intelligence. The contents of intelligence are only given through experience, in the perceptual relations with objects. But what we might understand as the form of intelligence, its typical styles of operation, Bergson regards as inherited, a tendency that needs to be used to be actualized but that is available for use in conscious beings. He hypothesizes that whereas instinct has an innate knowledge, enacted rather than conceptualized, of its objects, intelligence has the innate capacity to comprehend relations. It is significant that in characterizing this innate capacity, Bergson does not refer to an a priori knowledge; instead, he uses the example of language to illustrate that even if the entire contents of language must be learned, the facility for language is innate in humans. And this facility is dependent on certain intellectual properties: the capacity to generalize, to see resemblances, to make contiguities, to infer, to extend.[13]

For Bergson, what is innate is the capacity of intelligence to indefinitely extend itself through relations of similarity, containment, causation, attribution, and so on, an innate capacity to think spatially, so that once a term, concept, or language is understood, it can be used to generalize from what is known to what is unknown, it can presume a homogeneous space of generalization. The facility for language, to take the most significant example of this innate intelligence, is remarkable insofar as it can generate, from a finite number of sounds and letters, terms that apply to an infinity of objects, real or imagined.[14] What is innate is its access to and reliance on form, which is always generalization to the degree that it is able to be separated from matter. If instinct has an innate knowledge of the life that interests it, intelligence has an innate knowledge of form, "an external and empty knowledge" (CE 150), able to frame an infinity of objects, able to induce, generalize, universalize, to cut up the world, not according to the givenness of the object as it presents its effects through perception, but according to other (indeed, an infinity of other) criteria.

This formal or abstract knowledge of relations means that although intelligence is able to address any category of objects, it lacks any immediacy of contact with materiality, with any objects in particular. This does not mean that we should consider the knowledge produced by the intellect to be speculative or reflective. It is a knowledge primarily linked to action, to the highlighting of those characteristics of objects that will best facilitate our action, and to the obscuring of all others. It is a knowledge differentiated from the narrowly pragmatic impulse of instinct because, unlike instinct, its aim is construction or fabrication. This is why it finds itself most at home in inert matter, which lends itself readily to the processes of decomposition and recomposition necessary for making. Intelligence aims at fabricating objects, tools, machines from inert matter by reconstituting and reorganizing its parts in production. This is why it is more able to deal with matter than with life, with solids rather than forms of flux or flow; this is also why it feels at home in the spatial qualities of extension, which represent objects in external relations with each other. Intelligence is attracted to that which can be divided and recomposed according to our needs. Intelligence craves discontinuity, needs to cut up and divide the material world, seeks differentiation. It is for this reason that instinct, dormant as it is in intelligent beings, is more attuned to the flow of life than is intelligence, whose interest is more oriented to engineering, to the design, production, and use of clearly demarcated objects, to the fabrication of material according to an imposed form than it is to the unrepeatable singularity of each life or each moment in a life: "Life, not content with producing organisms, would fain give them as an appendage inorganic matter itself, converted into an immense organ by the industry of the living being. Such is the initial task it assigns to intelligence. That is why the intellect behaves as if it were fascinated by the contemplation of inert matter. It is life looking outward, putting itself outside itself, adopting the ways of unorganized nature in principle, in order to direct them in fact. Hence its bewilderment when it turns to the living" (CE 162).

Intelligence feels at home in organizing matter. Its primary interest is in facilitating action and invention, not in understanding them. Its natural culmination comes with the development of science, which attempts to produce invariable rules for the regulation and prediction of matter. The operations of logic, mathematics, and other forms of division and clarification conform to and provide the contemporary pinnacle of intelligent accomplishment. But intelligence, though infinitely perfectible, is nevertheless limited; in particular, Bergson claims, it is unable to think two central facets of the continuity of duration: the multiplicity of qualitatively different elements and their blurring interpenetration. It is unable to proceed without clear lines of demarcation, discontinuity, and boundaries, without unambiguous forms of identity. This means that intelligence is not adept in its dealings with lived continuity, experience, subjectivity, sociality, and all that has to do with life itself: it tends to impose on living bodies the qualities it seeks in inert matter; it tends to mechanism as its natural orientation: "The intellect is not made to think *evolution*, in the proper sense of the word—that is to say, the continuity of a change that is pure mobility" (CE 163).

In place of a multiplicity of elements and their blurring interpenetration, that is, in place of the endless becoming of life, the intellect puts homogeneous and immobile states; it positions stable things as the measurable base or substance for fleeting accidents; it sees change as a surface ripple on that which is fundamentally immobile; it reduces the mobile to the immobile. Its forms of knowledge are always views taken from the outside, an external breaking down and reconstitution. Bergson contrasts the instinctive knowledge of paralysis the wasp performs on its caterpillar prey with the knowledge of the entomologist: the insect knows, internally, by sympathy, perhaps through coevolution, where to sting to effect paralysis rather than death; the entomologist, by contrast, knows it only through experimentation, through learning where each nerve center is in the caterpillar, through adding one piece of information to another and building up the whole through aggregation (see CE 173).[15]

Intelligence is thus unable to deal with change, evolution, life, or duration. What is new escapes it, for it strives only to extend what it knows, not to question how it knows; intelligence projects onto the unknown what it has already confirmed, what is capable of being extended or elaborated, what has knowable consequences, what is able to be repeated, controlled, predicted. It applies itself to things, but only from the outside, reconstituting their parts but understanding nothing of the interpenetrating totality from which these parts are derived. It is unable to enter into life in its own terms.

To summarize, then, what the tendency to intelligence implies:

1. Intelligence is freed from the constraints of an instinctive activity: it no longer requires a preset behavioral response to a triggering stimulus, but is able to select among a range of possibilities and to forge altogether new reactions.
2. Intelligence is oriented primarily to the manipulation of unorganized matter and the spatialization it entails.
3. Its manipulation of unorganized matter results in the production of organized tools, instruments, which enable that manipulation to be facilitated with more agility, scope, and effectiveness than through instinctive activity.
4. Intelligence has an innate knowledge, not of the object, but of the kinds of relations that manifest themselves most readily in spatial terms: relations of similarity, contiguity, containment, causation, properties, adherences, and so on. As such, it has a fundamentally geometric orientation.
5. The knowledge intelligence acquires by learning and technological extension is thus an external knowledge of things, objects and subjects known from the outside, known insofar as some predictable or stable element can be extracted from them but not understood from within.
6. Its knowledge is oriented primarily to questions of quantity, degree, and magnitude rather than to the discernment and understanding of quality and intensive differentiation. It reduces quality to quantity, differences in kind to differences of degree.

Intuition

Life separated instinct from intelligence, not definitively but historically, in evolutionary terms, as two different directions that it could follow, tendencies, orientations, modes of knowledge and forms of behavior, each of which developed and elaborated itself with remarkable agility and variety in its own directions, while nevertheless retaining the most rudimentary connections with the other. Where intelligence oriented life outward, to matter, space, and objects, instinct oriented it inward, not through reflection or introspection, but through a kind of internal sympathy which directs it to act on, and for, the vital elements of life relevant to it. Instinct lives an internal existence, orienting life toward the creature's own survival and place among other members of its species, as well as other species relevant to its needs. Intelligence performs more and more prodigious feats in its manipulation and regulation of matter, in its transformation and reconstruction of the material world, which functions now as its prosthetic extension and its mirror reflection. Ironically, though instinct thrives on a certain degree of stability and relies on procedures developed in the past to propel it into the future, and though intelligence directs itself to the control of a foreseeable, regulatable future, instinct lives naturally in the particularity of duration, whereas intellect finds its most conducive milieu in the generalized form of spatial relations. Although instinct knows nothing of measurement and organization, it nevertheless fully lives in duration, places itself in the movement of time, subjects itself to temporal flow, even to evolutionary and phylogenetic metamorphosis. Intelligence knows about spatial relations, relations of simultaneity and contiguity, but it is unable to deal with the continuity and succession of duration, except through its tendency to spatialize duration and reduce it to measurable time. Each direction is propelled by the vital impulse of life; each has unlimited potential for elaboration and development; each constructs a kind of symbolism or language of its own; each goes as far as it can in its adequacy to its objects. There is no real external measure of their comprehensiveness and productivity, for each performs its task of supporting and elaborating forms of life in its own way.

Although they are two contrary movements, one (intelligence) directed outward, the other (instinct) directed inward, Bergson seeks a way of returning each to the other, or rather, of finding some principle between the two, which derives from intelligence its capacity for abstraction and generalization, and from instinct its sympathetic apprehension of and openness to life. He calls this knowledge intuition. Intuition is "disinterested, self-conscious" instinct (CE 176); equally, it can be understood as intelligence now attuned to itself and to the specificities of life. Akin to an aesthetic rather than a scientific understanding, intuition is the close, intimate, internal comprehension of and immersion in the durational qualities of life. Intuition is not an alternative to instinct and intelligence, but their orientation in different directions. It is the orientation of the rudiments of instinct with the insights of intelligence, no longer directed to a single or given practical end, but for its own sake. The observational and aesthetic appreciation of, say, a work of art is not the simplification of the work to its most recognizable features; it is an immersion in as much of the art object's qualities as one can achieve, not simply to learn something or do something but primarily to feel something, which may, but often does not, have a practical concern. This involves both a lessening of the intellect's grip on the object's future use and a deepening of its capacity to scour and address the multiplicity of its other (nonutilitarian) qualities. It is a contemplation or observation that opens up worlds to us, rather than narrowing the object down to our potential concerns.

As a mixture or combination of instinct and intelligence, intuition remains outside of either, though Bergson considers it the evolutionary heir of instinct more than intelligence. He regards intuition as the only mode of apprehension of differences in kind and of that virtuality that resides in the living transposition of the past into the present. Intuition is the only way we can consider duration and becoming without transforming them into representation and without considering them cinematographically. He appeals to philosophy, metaphysics, as an other to science, and intuition as the precise method and means by which philosophy is able to apprehend being and the world "from the inside."

Although it cannot be identified with duration, intuition is a movement by which thought emerges from and recognizes its own relation to duration and thus the relations of difference and entwinement it has with other durations. Duration is the reality in which intuition finds itself, but intuition is also the only mode of access to the singularity and generality that is duration: "To think intuitively is to think in duration. . . . The intuition we refer to . . . bears above all upon internal duration. It grasps succession which is not juxtaposition, a growth from within, the uninterrupted prolongation of the past into the present which is already blending into the future. It is the direct vision of the mind by the mind—nothing intervening, no refraction through the prism, one of whose facets is space and another, language" (CM 34–35).

Intuition is a method, a way of knowing, that bypasses the divisive impulses of intelligence. It is not to be confused with feelings, sympathy or empathy, or being in tune with, which give rise to instinct, but is a quite precise method, capable of being developed and honed, for apprehending duration. Just as intelligence has a fundamentally practical orientation and instinct is an unreflective enactment of an unconscious or latent knowledge, so intuition is a fundamentally reflective orientation, which is directed to the apprehension of the whole, a whole that exists only in duration, as becoming. Because intelligence divides the world according to the needs of action, to diminish the richness of the object in order to facilitate selective action, intuition must be regarded as a more complete, less self-invested mode of access to and immersion in objects and the world directly, internally, without mediation. Intuition returns to the real the fullness and interconnectedness that intelligence subtracts from it.

The intellect approaches ever more closely the constitution of matter without being able to grasp it from within. This is what leads Bergson to describe it as cinematographic: it produces snapshots, perspectives, on the object, which, however much they are magnified and proliferated, still never manage to capture the specificity and details of the object:[16] "For intuition the essential is change: as for the thing, as intelligence understands it, it is a cutting which has been made out of the becoming and set up by our mind as a substitute for the whole . . . Intuition, bound up to a duration, which is growth, perceives in it an uninterrupted continuity of unforeseen novelty" (*CM* 39).

Bergson insists that life is always broader than knowledge, and the intellect should be regarded as only one stream of life, a stream that doesn't distinguish man from animals, but is their shared evolutionary inheritance. If the intellect is formed from its dealings with inert matter, by contrast, intuition, derived more directly from instinct than intelligence, is directed both inward, to grasp oneself and one's being as an internality, and also outward, insofar as it directs us to the unique particularity of objects. Intuition is the evolutionary heir of instinct, its refinement through its intermingling with intelligence.

If intelligence culminates in science and its technological accompaniments, intuition is the proper terrain of philosophy, whose interest is never directly practical but always mediates action through the construction and elaboration of concepts. Rather than being anti-intellectual, an abandonment of the principles of reason, as Bertrand Russell claimed, intuition is a refinement of the clear-cut divisions of the intellect, a complication of their relations of mutual support, and the addition or return of what these divisions leave out. Intuition affirms the absolute specificity of its objects, whereas intellect can only seek out generalities, resemblances, and similarities. Where intelligence seeks ready-made categories and concepts by which to clothe real relations, intuition seeks concepts uniquely tailored to the object alone, a concept cut out specifically for each "thing" intuited.

Because intuition can function only if the operations of intelligence are temporarily cast aside, that is, if the search for means and ends is delayed, it can occur only with great effort and for short periods. It functions intermittently, in short bursts of insight and connection. It is immensely difficult to sustain, for we must return to the pragmatic requirements for which the intellect has prepared us. The effort to generate intuition is "arduous" and cannot last (*CM* 39).[17] It occurs not so much through attention as through a leap into the movement of what is new, a leap outside the familiarity that intellect provides, into the unfamiliar, where we must seek out new means of apprehension. This is partly why, even with the effort of attention, intuition always begins with indeterminacy and vagueness and is shadowy in its origin. The objects to which it is directed cannot be readily comprehended, for intellect decomposes the new and the mobile into the already known and the immobile. What is apprehended by intuition requires the mediation of intellect, language, representation to be crystallized, communicated, to become a form of knowledge, for intuition has no capacity to represent itself except through its intellectual extension. Intuition thus occurs only through a provisional incomprehension, and through patience, a process of seeking the precise forms of concept, method, and representation that suits what is intuited in its simple immediacy. Thomas Kuhn (1970; see also De Landa 2002) understood this as the moment of insight in scientific discovery that shifts a paradigm. Paradigm shifts in science and in other forms of intellectual innovation are the result of a growing, developing intuition of something outside existing frames of reference, dimly perceived at first, which threatens to transform already given or accepted paradigms or to abandon them, to make models newly relevant if they succeed in drawing to themselves some of the apparatuses of scientific method, if, that is, the intuition can be recast in the (sometimes new) language of science.

For Bergson, intuition is a method or technique that precedes symbolization and representation but that requires them for it to have an effect. It is not a psychological process or skill, nor a personal accomplishment, but one of the biological contingencies that mark all of life, a tendency, more or less active or dormant, whose function is not synthesis but an acknowledgment of a mode of belonging to, immersion in, being part of a larger whole, a whole that cannot be totalized in a single force but that coheres through its conformity to the principles of differences and divergence. Intuition is a trajectory that generates a variety of forms, a variety of modes of sympathy, inner apprehension, or coincidence. It is a mode of direct contact or community with the object, a provisional coincidence with it that precludes projection, mastery, or judgment: "We call intuition here the sympathy by which one is transported into the interior of an object in order to coincide with what there is unique and consequently inexpressible in it" (*CM* 190).

Instead of sympathy or identification, which is nothing but a psychologization or subjectivization of knowledge, Bergson wants to link intuition to an understanding of the absolute, the domain of metaphysics. What the intellect provides is a relative knowledge, a knowledge of things from a distance and thus from a perspective of self interest (which is all that Nietzsche claims any knowledge is capable of), a knowledge mediated by symbols, representations, measurements, whereas intuition can provide an absolute analysis, which means one that is both simple and internal. This absolute is not understood in terms of an eternal or unchanging essence, but is, from the outside, a complex interplay of multiple forces and factors which, from the inside, resolves itself into a simple unity: "Seen from within, an absolute is then a simple thing; but considered from without, that is to say relative to something else, it becomes, with relation to those signs which express it, the piece of gold for which one can never make up the change" (CM 190).

What Bergson is advocating, in fact, is a refined and newly considered empiricism, an empiricism that avoids the models provided by Locke and Hume, for whom all relations are external, connecting otherwise unconnected objects, insofar as it refuses induction and abduction, the generation of general rules from the multiplication of singular instances, the generation of a system of probability that involves the repetition of distinct instances. Bergson is interested in singularity, in the absolute specificity of objects, rather than, as the British empiricists have assumed, the universal categorizations that emerge from groupings of objects. Intuition, unlike intellect, is thoroughly attuned to history, that is, to the undulations and nuances of time and contingency, to differences in kind. This is not a resistance to science, but rather, a plea for a more direct and unmediated relation to objects and to the world of matter, an augmentation of science. For Bergson,

> a true empiricism is one which purposes to keep as close to the original itself as possible, to probe more deeply into its life, and by a kind of spiritual *auscultation*, to feel its soul palpitate; and this true empiricism is the real metaphysics. The work is one of extreme difficulty, because not one of the ready-made conceptions that thought uses for its daily operations can be of any use here . . . an empiricism worthy of the name, an empiricism which works only according to measure, sees itself obliged to make an absolutely new effort for each new object it studies. It cuts for the object a concept appropriate to the object alone, a concept one can barely say is still a concept, since it applies only to that one thing. This empiricism does not proceed by combining ideas one already finds in stock . . . but the representation to which it leads us is, on the contrary, a simple, unique representation. (CM 207)

This other mode of empiricism is one that operates underneath the cuts, the division and discontinuities that intelligence imposes on the world, that calls attention to the fundamental interconnectedness of all "things," the fact that things do not occur in isolation from other things but are bound together in a continually changing series of streams that form a dynamic and continuous whole. Intuition is a mode of momentary, fleeting, and difficult-to-sustain access to this movement of continuity or flow.

In short, if, as Bergson says, "the intellect is characterized by a natural inability to comprehend life" (*CE* 165), it is only intuition that exhibits the alternative movement, not of disconnecting and dividing, but of binding and bringing together, of acknowledging change as fundamental and cohesion and continuity as integral. Intuition is the mode of proper comprehension of duration, and thus of life itself. This may be why Bergson insists that we all have an everyday access to at least some elements or forms of intuition, from those that stem in the first instance from the inner continuity that marks subjectivity and actions and the sensations and affections within subjectivity. It is this inner efficacy that I perceive without, in the world itself, where those things are also always interconnected and mutually defining, where changes in one lead imperceptibly to changes in all.

Although it is of course possible to resolve our own inner identity into a series of elements, as analysis provides, nevertheless, this experience of continuity and cohesion is itself difficult to evoke but pervasive and undeniable: "What I find beneath these clear-cut crystals and this superficial congelation is a continuity of flow comparable to no other flowing I have ever seen. It is a succession of states each one of which announces what follows and contains what precedes. Strictly speaking they do not constitute multiple states until I have already got beyond them, and turn around to observe their trail. While I was experiencing them they were so solidly organized, so profoundly animated with a common life, that I could never have said where any one of them finished or the next one began. In reality, none of them do begin or end; they dove-tail into one another" (*CM* 192).

The indeterminacy of identity and the pervasive overwhelming of being by becoming in inner or psychical life is belied by any psychology that focuses on psychical states or distinct processes, for these are only ever abstractions, disconnected from the flow of time and the coloration they receive from the whole of one's life history and from one's current experience. This means that, in effect, Bergson claims that there are two ways to know: one, intellectual, immobilizes and isolates, facilitates practical action and use but thereby moves from the real to its schematization; the other, intuitive, seeks continuity, indiscernibility, flow, and duration, immobilizes practical action but brings us directly into connection with the dynamism of the real. The first reveals itself most clearly in its manipulations of matter; the second expresses itself most directly in the subject's own inner cohesion.[18]

Bergson holds that the physics and metaphysics of inert matter are close precisely because the intellect is bent in the same direction as unorganized matter. This means that the scientific/intellectual approach to inert matter is "in tune" with the actual nature of this matter, tending as it does toward mechanism. Problems arise when this mechanistic approach is carried over to the study of the living. To repeat, Bergson does not locate the mistake within science; indeed, he remarks that science must approach living matter in this way if it is to make practical use of it. Of the "truth" offered by science as regards a living object, however, Bergson writes: "[It] . . . becomes altogether relative to our faculty of action. It is no more than a symbolic verity. It cannot have the same value as the physical verity, being only an extension of physics to an object which we are *a priori* agreed to look at only in its external aspect. The duty of philosophy should be to intervene here actively, to examine the living without any reservation as to practical utility, by freeing itself from forms and habits that are strictly intellectual. Its own special object is to speculate, that is to say, to see" (CE 198).

Science, the product of intellect, alongside intuition, a labored and difficult resonance with the duration of events, are two different lines of development and elaboration of the relations between subject and object, memory and matter. There is no question, for Bergson, that intellect has a primacy over intuition, not only historically and politically, but of necessity: intellect needs to acquire a certain level of proficiency, to be able to take care of basic needs, to distance itself from its objects, before the more luxurious and excessive operations of intuition are possible. But intuition, an attunement to the particular which thereby reveals its relation to the whole, is the compensation for what intellect must leave out, the myriad connections, entwinements, and transformations that make up even the most stable objects of intellectual analysis.

Intuition is not an exploration of the unknown (this is the task of intellect, to render the unknown known or knowable), but a finding of oneself in the unknown, an immersion in its specificity, a negotiation with its newness. It can be considered a mode of being lost in an unfamiliar element to which one must adjust through an understanding from within. Intuition, then, is the way the inner directedness of instinct can rejoin the outward orientation of intellect, which have been elaborated by evolution in opposite directions. Intuition is not the reconciliation of the contrary impulses of instinct and intellect; it is the generation of a new series of impulses which may help modify our relations to the world.

Unpredictable Futures

Bergson's works have languished largely in obscurity for well over half a century. Whatever their inadequacies—and any discourse or theoretical system, as he acknowledges, must have limitations and inconsistencies, must be subject to its own becomings and elaborations—they nevertheless remain of dynamic relevance to understanding the ontology and politics of temporality. Although his interests remain directed to the general metaphysical abstractions concerned with the complex concepts of life, matter, and becoming, his work is not without very significant implications for how we might reconsider the preeminent but unrecognized discourses of temporality—history, sociology, psychology, politics, and so on—as disciplines and methods of analysis bound up with duration. These disciplines have embraced, perhaps without knowing it, a dynamic and unpredictable temporality (retrospective analysis, the seeking of broad continuities and discontinuities), even while they develop techniques (atomization, quantification, measurement, statistical generalization) that attempt to reduce the irreducible temporal dimension and framing of their objects of analyses into spatial representations (historical periods, lineages, causal chains, etc.). Duration is forgotten, although, ironically, forgetting is a continuing expression of life's adhesion to durational movement, is the condition of life's emersion in time.

Bergson continues a tradition—perhaps it could be considered a countertradition—that begins as early as the ancient postulate of continuous change but emerges in its contemporary forms through the explosion of Darwinist science, a science that no longer sought determinate forms and predictable relations, but turned to seeking out continuity and change, that reveled in the indeterminacy and contingency of its objects, that refused, at least at its outset, to translate the categories of biology into the language of chemistry or physics. In Darwin's hands, evolutionary biology, along with linguistics and economics, became conjectural rather than hard-core sciences, sciences linked to and embedded in history, sciences (if they are that) that attain whatever certainty they seek only when their objects of analysis have ceased, only in retrospect. Bergson points out that the persistence of the past in the present is the condition of all life and, by implication, of all politics, all social life, all cultural activity, and ultimately, insofar as we are able to conceive of material existence in its rich interconnectedness, all material life as well. Nietzsche revealed that our containment within and as historical beings is not only the mark of our activity but is also the possibility of our redemption, our self-overcoming. The privilege of an open and unknowable future liberates even as it frustrates our aspirations, goals, and ideals.

Insofar as time, history, change, the future need to be reconsidered following Bergson's understanding of duration, concepts of the virtual and of differences in kind may prove central in reinvigorating how we conceptualize the future and engender it through political and social action. He refuses to tie the future to the realization of possibilities (the following of a plan), linking it instead to the unpredictable, uncertain, diverging, often impromptu processes of actualization. The emergence of the future is not that which the possible, a plan, blueprint, or algorithm, prepares us for. It is not a step-by-step process, reproducible through computation (computational simulations are fundamentally reconstructive projects). There seems to be a confusion between the simulation of life, for which we can plan and prepare and which we can program, and life itself, which always unfolds without a ready-made script, with no program, goal, or aim, though it can be retrospectively interpreted as following a plan or path.

This distinguishes the Bergsonian conception of time and the future from much fascinating work in, for example, the proliferation of intriguing and boundary-transforming research on the simulation of life,[19] the rendering porous of the boundaries between life and nonlife at the Sante Fe Institute,[20] and the recent provocations of Stephen Wolfram in *A New Kind of Science* (2002), who claims that life, matter, the universe, and all its contents function according to algorithmic or computational programs of great simplicity to generate immense complexity. One would not want to disagree with the leap forward from static mathematical equations to dynamic algorithmic recipes; but it must be made clear that, from a Bergsonian perspective, algorithmic models share the same philosophical or ontological problems as mathematical models. Neither affords duration any real existence, accepts it as a force of its own, sees it as a positivity, as a productive and counterbalancing force to the positivity of space. Algorithms, as much as equations and formulae, could speed up or slow down the duration of their steps and stages without in any way affecting their results. Time becomes merely the neutral, regulatable background in which objects or relations change, rather than an inherent ingredient in such research. This research on the simulation of various forms or practices of life is both intriguing and worthwhile, but what it tells us about life itself, and about the very real immersion of life, whether of animal species or of social and economic relations, in the uncontainable and uncontrollable movement forward of time is considerably less than it proposes. The mechanical representation of life, however complex these mechanisms are, always remains a representation of a certain element of life, certain extractable facets of life, rather than its interconnected continuity.

Evolution is an ongoing experiment, not in the scientific sense of controlled variables in a closed system, but in terms of continuing variations, risks, explorations, only some of which will develop themselves and many of which will perish in their elaboration. Politics is, in this sense, evolutionary too: political struggles and encounters are not simply the outcome of random chance, but a clash of prevailing forms with those experiments in otherness, in difference, which challenge their range, scope, or effect. Like evolution, politics has no telos or pregiven goal, no preferred methods, no privileged terrain: it functions, like all living things, strategically, utilizing what of itself and its surroundings it can to do what it believes it is able to. The conclusion turns more explicitly to the question of politics and how reconceptualizing time may help transform how we understand politics.

CONCLUSION: THE FUTURE

Adapting is not repeating, but replying.—Henri Bergson, *Creative Evolution*

This book has focused on the reality of time, its relevance to understanding natural and cultural systems, its resistance to and otherness from spatial representation, and its irreducibility to other terms and frames of explanation. Against the prevailing impulses of mainstream philosophy and conventional physics, I have utilized the work of misunderstood, wayward, or neglected figures who have refused to see time as linear or countable, as the fourth dimension of a space-time continuum, as a neutral or passive medium for the highlighting of objects.Instead, I have theorized time as active, positive, with its own characteristics, features, and specific effects, which remain difficult to ascertain and impossible to measure insofar as "empty time," "pure time," and "time in itself" are not cognizable by or comprehensible to us. Time's capacity to hide within objects, through and as things in time, means that to the extent that we focus on the nature of objects, we obscure the nature of temporality. Although the evidence of its effects appears only on and as these objects in time, objects in their totality (through their various irreversible processes) and living beings in their particularity (through their development, aging, and mortality) attest in a fundamental manner to its irresistible force.

Knowledge and the Future

Although time is *experienced* in the various temporal rhythms and processes that mark the living being, it cannot be reduced to subjective projection alone, to the domain of the psychological, an illusion life imposes on the temporally indifferent universe of matter. Newtonian and Einsteinian laws posit a reversible time, a temporality that could as readily flow backward as forward.[1] Yet scientific laws are less objective and more pragmatic in their operations than they may seem to their practitioners: we must understand that scientific laws are in fact the formulation of very precisely delimited generalizations, principles that make the material world comprehensible to and useful for human observers and agents under highly controlled circumstances. They are not moral or enforceable laws, inevitable and irresistible, but shorthand descriptions of perceived regularities that are of use to us, descriptions that many scientists (since Maxwell and Boltzman) believe are statistical probabilities rather than universal laws. Such statistical analyses, while telling us much about general patterns, tell us very little about any particular or concrete entity. Scientific laws are the expression of our lived need or desire for regularity, for repeatability, for a sense of control over and security in a universe larger than our comprehension or control. The sciences have been spectacularly successful in generating technologies that produce certain regularities, even if their explanatory power has left much of the universe—particularly its irregularities, the events that transform it, its qualitative metamorphoses—unexplained or enigmatic. Science helps facilitate our practical use of objects in the world by delimiting and analyzing various material processes and activities, but it has been unable to adequately address that which frames the practical, the interconnections between units and systems, the ways that systems are cut out from a cloth that unites and intermingles them with all other systems and processes in fundamental and unbroken continuity with them. To understand material systems, including their ramifications in and as cultural systems, that is, in their interconnectedness, one cannot simply add together the various well-analyzed, independently conceived subsystems. Their integration is not additive but transformative: it is performed not only through spatial relations but above all in temporal relations. It is not only living systems that evolve in time, but worlds, universes.

The divergent line that runs from Darwin through Nietzsche to Bergson (and beyond, to Deleuze and Irigaray) explored in the previous pages reveals that time is not only the regulative force of life as we know it (as Darwin claimed); it is the very motor of the universe as a whole (as Bergson implied), as well as the principle of life to come, life that overcomes itself (as Nietzsche affirmed). Time is in principle outside, before, and beyond matter, a precondition of matter's emergence, and the force that, surprisingly, without predictability, rends life from its more unstable interactions. Time imposes on matter the task of extending itself, of elaboration; as Bergson suggests, it is only the movement of time that ensures that not everything is given at once, that the universe is not over the moment it is born, that it occupies a duration, whose rhythms and speed, if they changed, would alter everything. Time is protraction, not just the hesitation of the living as they open up new possibilities from those virtual to existence, but of matter itself as it becomes more organized and its interactions more expansive, complex, and unpredictable. Life protracts the temporal delay latent in physical processes into a productive freedom, an indeterminacy, into the creation of the new, into invention itself.

The scientific inclination to the reduction of time to modes of the spatial and the countable has left science with a number of paradoxes, potentially insoluble problems, problems that science cannot in the long run ignore and that threaten to transform the nature of the sciences themselves as they continue to be drawn again and again to these inherent and irresolvable problems: How can science explain the origins of the material universe if time and space are the contingent, emergent properties of matter rather than the conditions of matter's emergence? What is time (or space) before there is matter, before the eruption of the Big Bang? And if time is nothing, then how can the Big Bang come about and what terms can we possibly use to explain its initial rapid inflation? How can our universe be understood as temporally and spatially finite, with a beginning and an end, if there is no reality to eternity, no structure in which time's limit has any meaning? What is the force of time in dissipative systems, in physical systems (such as chemical reactions) that seem to exhibit temporal irreversibility? Are such systems explicable in terms consistent with the basic principles regulating units in the physical world? And if nonorganic systems are temporally reversible, yet organic systems are time-sensitive and irreversible, how is any integration of physics into biological models justified without fundamental transformations in both disciplines? In the long run, these are questions science must be able to address in its own terms and using its own methods. They abide, they are questions that will not disappear whatever conceptualization (or lack of it) we develop. The question of time persists, in spite of our efforts to dissolve it in calculation and quantification; it continues to press on us, and all our creations, including science, with no relief.

These big questions are at the very threshold between philosophy and science, where both cannot but touch on ontology. They are questions that the sciences must explore in their own way, that may—perhaps even must, in the long run—change the questions, orientations, and methods that direct the sciences themselves. They are the region where science participates in metaphysics, where it encroaches on the durational questions that occupy philosophy. Philosophy has an opportunity to return thought to, remind thought of, what the sciences must leave out, the discernment of what integrates natural systems, material or cultural, as well as the temporal processes, syntheses, elaborations that enable both integration and conflict to be possible and to exercise their evolutionary effects. Philosophy may remind us that there are other possible forms of science, different ways of knowing, than those that prevail at any moment. In this way, philosophy may ally itself with feminist theory, which insists on the sexual difference not only of bodies and their pleasures, but also of perceptions, knowledges, systems of representation, and conceptual models. Feminism reminds us of the forgotten, lost, or virtual history of the contributions of the "others" of Western reason and how knowledges represent many interests and generate many perspectives, only some of which become privileged, funded, and dominant. Philosophy may remind us of the rich potential of intuition, the impact of affect, the force of intensity that such dominant models of knowledges leave out.

Evolutionary models, as I have stressed throughout, are powerful resources for understanding the emergence of complexity. Through the processes of reproduction and natural selection, sometimes extremely simple organisms, processes, or activities evolve into remarkably complex and unpredictable patterns, creatures, organizations, strategies, and creations. The interaction of Darwin's two principles of biological evolution—the heritability of variation and the operation of natural selection—if they are relevant at all, have as much explanatory power in social and cultural systems as they do in natural and organic systems. It seems plausible to suggest that economic, linguistic, political, and conceptual systems emerge and develop according to a similar temporal logic, although their material forms and spatial arrangements vary widely. Rather than operating as consciously regulated systems, the product and effect of their agents, or else as deterministic, pregiven systems whose future is already contained within their present, cultural and natural networks remain open-ended and exhibit emergent properties as they increase in complexity over time. They face events, transformations, challenges, upheavals, as does the natural world. Such systems are time-sensitive: they grow more rather than less complex over time; they develop unexpected properties or qualities not given in their past; and as time moves forward, their characteristics are capable of major upheaval and realignment. In short, social, cultural, political, artistic, and intellectual institutions and practices also evolve unpredictably, without plan, without overarching control; they can be considered the emergent patterns that result when large numbers of individuals and collectives undertake social and cultural as well as biological activities, which come to resonate and interact with each other, to form planes of consistency, with ever greater complexity. The emergence of complexity, whether on the scale of the biological (cells, organs, individuals, collectivities, etc.), the natural (weather patterns, wave forms, chemical oscillations, vortex formations, etc.), the cultural (economic fluctuations, conceptual debates, the transformation of languages, or even the flow of traffic), requires both a conception of temporality that privileges the future over the past and maintains its relative autonomy from the present, that is, some concept of temporal irreversibility, and a nonlinear conception of structure, a structure describable primarily in terms of the incalculable movements of small units in large numbers, that is, in terms of statistics, in which small variations at the level of individuals can produce large-scale organizational transformations. In other words, an open structure.[2]

Such systems, natural and social, have been described as "on the edge of chaos," as "nonequilibrium processes" or "dissipative structures" (Prigogine 1997; Prigogine and Stengers 1984). While various physical systems may function according to Newtonian principles when they operate under conditions of equilibrium or stability, Prigogine claims that, on the edge of chaos, where such systems undergo unpredictable modifications and developments, where they enhance themselves through phase transitions that complicate their organizational structure, they can be understood as close to life itself, approximating the surprise for the living of the unexpected and the unforeseeable.[3] Such systems are particularly sensitive to temporal movement, and though they may form relatively closed systems, the sheer volume of units, the vast number of individuals they involve makes their organizational structure, as well as these individuals, mutate, develop, complexify. These systems, and their elements, whether cultural or natural, can be said to evolve, to become, to undergo irreversible self-transformation, and to provide the context in which their individual members evolve and develop. The ability to become, that is, to increasingly exhibit emergent properties, cannot simply be confined to life itself, for it is the condition under which nonorganic configurations of matter give rise to the earliest forms of life, as well as the condition under which life elaborates and develops itself, along the lines matter itself provides, into the complex, divergent, and unexpected directions—among them, sexual bifurcation—it has taken and continues to take.

Time is real, its characteristics are unique, its effects cannot be explained in other terms. It distinguishes itself from space, from objects, from its multiplicitous representations in mathematical, formulaic, or geometrical terms, or in the images and representations provided in the visual arts, through effects that are not spatial, objective, measurable, or quantifiable, although it has no language of its own, no models on which to base itself except those provided by the impulse to spatialization. Indeed, the very notion of a model entails spatialization and resists duration; space is the ongoing metaphoric site for the representation of temporal qualities. Yet, the "logic" of temporality does not involve static self-identical terms, terms that can be laid side by side and evaluated or compared, but those more appropriate to the self-changing, that which has no self-identity but continually differs from itself. It has its own durational "rationality," its own qualities and qualitative effects. The reality of time is not reducible to the reality of space and of objects, though it can only be seen in its effects on them. It is most directly understood and experienced outside of images, models, and representations as a force, an impulse forward that cannot be resisted, for it is lived as the subject's own growing, waiting, reminiscent, and anticipatory continuity. Among the more significant of these temporal qualities and effects are the following.

1. Time lends itself to spatialization, enumeration, and geometrical modeling, although these processes lose what is essential to temporality: its dynamic movement. Time thus enables counting (as Bergson demonstrated, there can be no concept of number without synthetic capacities that are temporal, for each number represents both an individual and a synthesized collection or multiplicity) but changes its quality when counted. It exhibits a passivity or compliance relative to the models, images, and representations by which it is understood (this is why it is immune to scientific testing, to empirical verification or falsification), demonstrating none of the resistance that material systems can exert over (mistaken or partial) representations. Nevertheless, ineluctably, relentlessly, it exerts its own active force of aging, recontextualizing, or overcoming. (Whatever scientific or artistic representation we presently believe "truthfully" captures some features of the real, we can be sure that, with the passage of time and with enough interest in the question, these representations will be superseded.) It works its effects on those very discourses that deny it!

2. Time does not operate through relations of containment, contiguity, or

coordination, though its spatialized modeling provides us with the prevailing metaphors and a language of the spatial and the geometrical by which to represent temporal relations. It is only through such spatializing models and metaphors that one time contains another, can be construed as larger or smaller than another, that begins when another ends. Time is a continuity, though the events in time each have a duration of their own, and thus function through discontinuity, realignment, or rupture. Time is what enables events, in retrospect, to be linked to their antecedents and consequences, to spread out their implications and effects. Events may be conceptually decomposable into their elements or ingredients, but only retrospectively, and only at the cost of understanding their unique and unrepeatable interrelations.

Time can be understood as always doubled. There is a cosmological force: time as a whole, time without regard for objects, time in itself. It is only in relation to this cosmological principle that events can be understood as taking time, having duration, having temporal relations—before, after, simultaneous—with each other. It is only because of this cosmological ordering that events can be positioned, though not without some difficulty, relative to each other, given a chronology, a temporal location, a historical position. But to focus only on the ordering of events, their location in a measurable, overarching time, is to ignore the specificity of the duration of each event, its own unique temporality, the time of each thing or process. So these two times do not really contain each other: the cosmological is only "larger" than the event through a notion of magnitude alien to temporality. Instead, cosmological temporality subsists or inheres in the temporality of unique events; it is their virtual background or context.

3. Although the present does not contain the past, they coexist. They function simultaneously. This coexistence of present and past, the ways the past grows and augments itself with every present, the virtual potential the past brings to each present, provides it with a capacity to enrich the present through resonances that are not themselves present. Time doubles itself not just on a cosmological/individual scale; it also doubles itself at each moment of its continuity. The present and the past run simultaneously. This splitting of the present, at each moment as it occurs, into the virtual and the actual, the past and the present, creates the possibility of temporal resonances, of inventions and productions impelled by the force of the past filling in the details of the present, and the capacity of the present to reenervate elements of the past.

4. This coexistence of past and presents entails, not two parallel temporal paths, but the filling in of one (the present, the perceptual) by the richness of

the other (the past, memory). This might be seen in the overriding significance of a so-called founder's effect. Because time cannot be erased, and because the past is beyond the control of the present, the present necessarily carries some of the germs of the past as its present limit and as the future's capacity to differ from it. The historical force of what was, however rational or irrational its foundation, however steeped in violence or coercion, has an ongoing effect on the present. This is as true of the history of living organisms as it is of economic and political history. To be more concrete: the forms of life, or forms of politics, that we know today are not rational choices or decisions that biology or political consciousness once made. They are contingent forms, developed randomly, through mutation and accident, through luck or force. Whether or not they are evolutionarily successful, their justification is not that they are the most worthy, the best designed, the most tested, but simply that they live; they continue an arbitrary trajectory in the best ways that they can. Historical processes cannot operate without such a contingent foundation: what was helps to prestructure what is, although it also contains the conditions for what differs from what is. These present traces of the past cannot be erased, simply put to work, or acted, differently.

5. The present itself, always a continuous present that never passes into the past, is nevertheless not present to itself. It remains fractured and refracted through reminiscence and anticipation, the murmurs of the past and the potential of future. Every present is riven by memory, by the recontextualization of the virtual past which provides the links, contexts, and framework to augment the present's performance of perception, action, and reflection; equally, the present always spreads itself out to the imminent future, that future a moment ahead for which the present prepares itself by reactivating the past in its most immediate and active forms, as habit, recognition, understanding. The present acts, is active, makes. But what it makes is never self-identical, stable, given as such to the next moment. What it acts or makes is the condition of the transformative effects of the future. The present not only acts, it also consolidates, the past; it doubles itself as both present and past, actual and virtual. And it is only this doubling that enables it to resonate with the resources, the virtual, that the past endows to it and to the future.

6. The virtual exists only in time. Objects in space carry all they require of the past within themselves as their present. They are actual, manifest. The past subsists in the present only for living beings or for open systems: it is only for living beings that history is significant, that the past can modify the

future and the future the past, providing possibilities that the present hasn't contained. The virtual is the condition of being otherwise than what something is at this moment, its capacity for self-modification, elaboration, overcoming. It is thus the very possibility of the cultural, the political, the conceptual. The virtual is another name for the inherence of the past in the present, for the capacity to become other. As Nietzsche recognized, the weight of the past is also its lightness, its gift. Though we cannot change the past, we can use it to change the present. It is the condition of innovation and the new: the new can be formed only through a kind of eruption or interruption of the present that does not come simply as a gift from the future but is a reworking of the past so that the present is different from itself, is open to an eruption. The virtual is the resonance of potential that ladens the present as more than itself, that disrupts the continuity of the present, to open up a nick or crack, the untimely, the unexpected, that welcomes the new, whether a new organism, organ, or function, a new strategy, a new sensation, or a new technological invention.

These are highly abstract reflections and hypotheses. Such is the luxury of ontological conjecture, but such reflections should not be understood merely as a frivolous indulgence or without concrete implications. The relations between abstract speculation and concrete practice are always indirect, never reducible to simple application, and the outcome of such reflection is never clear or guaranteed. Nevertheless, concrete practices and pragmatic questions are themselves always invested in abstract and ontological assumptions whose transformation may help redirect, refine, or complexify those practices. My concern throughout has not been only with the scientific implications of reconsidering temporality in its own terms, but more with its cultural and political consequences. A reconceptualization of time as continuity, as the order of transformation and change, as the condition of emergence and the active force of complication may provide a different angle from which to engage in political and cultural analysis and action.

Politics and the Future

I suggested in previous chapters that one of the implications of regarding time as a positivity is the confirmation of the ongoing significance, the virtual relevance of the past in refiguring the prevailing forces of the present. The resources for overcoming forms of domination, coercion, oppression, that is, for producing a new set of social and political relations and new values in culture, comes only from the excessive productions of the past, the virtual force that lies still immanent in but undeveloped by the present, the dislocations between past and present that make room for unpredicted futures. The past produces the resources for multiple futures, for open pathways, for indeterminable consequences, as well as for those regularities and norms that currently prevail. The present, with its structures of domination, has actualized elements, fragments of the past, while rendering the rest dormant, inactive, virtual. This means that the future, possible futures have the inexhaustible resources of the past, of that realm of the past still untouched by the present, to bring about a critical response to the present and ideally to replace it with what is better in the future. The past persists, and its virtual contents are rearranged, restructured, with each passing moment; it is because the past is contracted throughout the continuity of the present that history remains a political force, the site for the unraveling of the givenness of the present.

What is this persistence of the past in the present? And how does it help us to rethink questions of politics, which are ultimately questions of change, of desirable futures? Living beings are the actualization of only part of the past. Indeed, any actualization leaves part of the virtual unactualized, and in the various processes of divergence and proliferation it also induces new virtualities, new lines of divergence. It is precisely this unactualized potential of the virtual that is the condition of all radical politics, which takes as its aim the transformation of the present. Take feminism as one example of this radical impulse in politics: feminism would not be possible, could not emerge or have any effects unless it, along with its patriarchal objects of critical analysis and overcoming, were equally the effects, the actualizations or consequences of patriarchal power relations. Virtual in these relations, which posit and produce the privileging of what is associated with masculinity over what is associated with femininity, is the movement by which the conditions of this privilege are revealed and the privilege itself is shifted. In other words, to be an effective political force, feminism (or antiracist or postcolonialist struggles) must be latent in and a product of the very systems or structures it challenges. Feminism is the virtual effect of patriarchy as much as male privilege itself, the mark of the dual or simultaneous contingency of all political regimes and orders and their historical force or effect. The history of patriarchy (or racism or colonialism), in fact, the history of patriarchy and feminism's intertwining and sometimes contrary impulses, produces male domination (the privileging of bourgeois Eurocentrism, global capitalism). It also produces, indeed entails, the rise of (various forms of more or less effective, historically contingent) resistance.

History produces not only the forces of domination but also the forces of resistance that press up against and are often the objects of such domination. Which is another way of saying that history, the past, is larger than the present, and is the ever-growing and ongoing possibility of resistance to the present's imposed values, the possibility of futures unlike the present, futures that resist and transform what dominates the present.

History, the persistence of the past in the present, and the ongoing revision of the past in the light of its changing relevance to the present, thus plays a dual role. On the one hand, it provides all the resources that systems of power, domination, subjection can command. No political regime, party, or group can gain domination without some ongoing movements of assent, affirmation, or reward from groups whose privilege in the past upholds and supports their present privileged status. This is the line of conservation in which the privileged groups in the present attempt to continue such privilege into the future. On the other hand, history also provides resources not utilized by dominant groups or movements, resources that are the conditions of effectivity of other groups, parties, or movements, a resistant line, a line that is not necessarily the recognized opposition or adversary to the dominant group (as opposing political parties often are) but what is uncontained by it. The past, in other words, is not only the condition of the present but also the condition of every possible future that may arise from or be made out of the present. Which is another way of saying that the past is infinitely reflective: it is revived, returned to relevance, rewritten (that is, actualized) in potentially infinite (future) forms. It is not exhausted by its present actualization and thus continues to provide the resources for each generation, each event, each future to follow. It is not so much revived by the present as it seeks activity, reactivation in whatever form the present may enable. The active force is not simply the present seeking out past resemblances and relevancies, precedents; it is also the past seeking to extend itself and its potential into the present, waiting for those present events that provide it with revivification.

Instead of the past being regarded as fixed, inert, given, unalterable, even if not knowable in its entirety, it must be regarded as inherently open to future rewritings, as never "full" enough, or present enough, to propel itself intact into the future. The past is never exhausted in its virtualities, insofar as it is always capable of giving rise to *another* reading, another context, another framework that will animate it in different ways. The past, in other words, is always already contained in the present, not as its cause or its pattern but as its latency, its virtuality, its potential for being otherwise. This is why the question of history remains a volatile one, not simply tied to getting the facts of the past sorted out and agreed on. It is about the production of *conceivable futures*, the future understood not as that which is similarly contained in the present, but rather, as what diverges from the present, what produces a new future, one uncontained by and unpredicted from within the present. This indeed is what I understand feminist politics—at least at its best—to be about: the production of futures for women that are uncontained by any of the models provided in the present. Rewriting, reinscribing the past is a way to activate these possible futures and, indeed, is their only political rationale. The past is not exhausted through its transcription in the present, because it is also the ongoing possibility (or virtuality) that makes *future* histories, the continuous writing of histories, necessary. History is made an inexhaustible enterprise only because of the ongoing movement of time, the pressure of futurity, the multiplicity of positions from which rewriting can and will occur. History is not the recovery of the truth of bodies or lives in the past; it is the engendering of new kinds of bodies and new kinds of lives. History is in part an index of our present preoccupations, but perhaps more interesting, the past is as rich as our futures allow.

If we view history as the long-term history of evolution, we see that the movement of evolution, its compressed presence in every living being, is the ongoing condition of all becoming. Life is a becoming beyond what it is because the past, not fixed in itself, never fixes or determines the present and future but underlies them, inheres in them, makes them rich in resources, and forces them to differ from themselves. All forms of life, even the most stable, develop, elaborate themselves, at very different rates, beyond themselves. But this evolutionary movement of self-transformation is not restricted to the slow, long-term movement of species and natural populations; it is also the regulating principle of the middle-scale movement of cultural and political events, linked to social convention and habit, and of the rapid movements of consciousness and subjectivity in an individual or between individuals. Life, whether biologically, culturally, or psychically considered, is this movement not only of the self-differentiated, but of a differentiation always directed to elaboration, complication, emergence—to excessiveness, growth, the forward pull of time.

Life, then, brings the resources of the past to bear on the present. These resources—habit, instinct, memory, learning—conditions for indeterminacy, for highlighting the present's useful qualities, are the conditions through which the present and its objects change their relevance, function in other causal chains and networks than they would without interference. Contestatory politics consists in those activities that affirm or problematize the values of the present in light of their past connections and future implications, in the ways the past opens up undeveloped possibilities, hints, intimations that may not have been elaborated, that nevertheless can be reactivated, animated in other contexts.

In a sense, then, life is always politics: it is always about the perseverance of one or many groups at the cost of others. But what has been victorious, that is, prevails at a particular period, does not wipe out the traces of all others, even those rendered extinct. The movement of evolution does not supersede that which is victorious and leave the rest to oblivion. The rest, the remainder, left out by dominant individuals, groups, species, are not simply the dead ends of history, its losers, what is left behind. What was once may still affect what will be, even though it may play no role in the force of what presently is.

Resistance, in whatever form it may emerge, is always latent in, the legacy of, the present, which may stifle certain clear-cut struggles but cannot contain the impulse to challenge it. The past provides the resources by which such challenges may be mounted, not directly, not through repeating what was successful in the past (there can be no repetition without an identity of circumstances, which, in historical events, is never possible)—that is, not through learning—but through its capacity to disrupt the present with forces the present has not actualized. The resources of the previously oppressed—of women under patriarchy, of slaves under slavery, of minorities under racism, colonialism, or nationalism, of workers under capitalism, and so on—are not lost or wiped out through the structures of domination that helped to define them: they are preserved somewhere, in the past itself, with effects and traces that can be animated in a number of different contexts and terms in the present. Identity struggles of all kinds—for women, for sexual, racial, or ethnic minorities—have profited immensely from the revisions of history that research into these oppressed groups has revealed, from the texts and documents unearthed, the practices and inventions discovered, rediscovered, or properly reattributed, that the inspiration and potential of understanding this as one's past, one's history, entails. What is it to be a woman or man, to be black or white, to be foreign or native, to be Jewish or Arab, to be one identity rather than another? Not the accomplishment of certain given skills, qualities, or attributes (in spite of the claims of adherents to identity politics, a politics forged on the identificatory relations of a subject with an identifiable group or number of groups), nor the attainment of something essential, given, or unchanging. Subjective identity, the belonging to social groups and categories, is always a matter of history, a history that may extend from one's own life story, through one's family genealogy, to cultural, political, and ultimately biological history, as broadly as one chooses. To be a woman or a Jew or black or working class is to take on, to make one's own, the history of what women, Jews, blacks, or the working class have been and have done; to see oneself in that history, but also to see in that history not only one's limitations in the present, but also the condition by which such limitations may be overcome.

Even more transgressive, perhaps, are those resources of the past not directly locatable in historical records, not reproducible as present representations, the heritage of past vulnerabilities, fragilities, fractures, and dislocations in power that reveal their effects in the present, to highlight and make clear the points of qualitative or intensive fragility in present power structures.[4] These are not elements of a history that leads only to oneself—the frame or "interpretation" that directs identity politics—but rather, a history of what is larger than oneself, a genealogy of the social and cultural itself (a genealogy that eventually, if taken back far enough, must return us to the world of nature, perhaps even, in Deleuze's terms, to a geology), a genealogy also of the events that imperil and transform the social and the cultural. The world of nature, the accumulated past of life itself, the past that all of life shares in common, is not what limits cultural and political activity and change, for no change would be possible without the lability, the openness to change, that the natural order has developed for itself. It is this openness that is the first condition of politics as much as of life.

The past is our resource for overcoming the present, for bringing about a future. The more we avail ourselves of its resources, the more enriched are the current possibilities of transformation. But, as Nietzsche recognized, an immersion in the past for its own sake can prove debilitating in the present. It is only a judicial and well-balanced history (not necessarily an objective, total, or complete history, but its opposite: a history balanced between a certain fidelity to the past and the demands of living in the present in the welcome anticipation of the future) that will provide the resources not only for a critique of present power alignments but also for their future reorganization.

The future erupts through a kind of leap or rupture—a phase transition, in the language of Prigogine, a moment of the eruption of the untimely or the nick—analogous to the leap into the past that constitutes memory proper. It is not the predictable, foreseeable continuation of the past. It is an unexpected shift, the shift produced by the unexpectedness of events, which reorients the past and whose reorientation or reanimation reorganizes its present effects without steps, in a continuity that is also a discontinuity, a becoming. This leap *is* politics as much as it characterizes life. Politics is not reducible to this leap, it *is* this leap: recognizing itself in the past that prepares it, all politics, from the conservative to the radical, aims to develop a future through some efforts of the present. The conservative aims to present this future as the logical development on and progress of the past, shepherded by the self-interest of the present. A radical line, depending on its degree of radicality, more or less accepts as desirable a dislocation or incoherence between the present and the future. The future as surprise is welcomed with more or less anxiety or openness. (Even leftist and feminist policy work self-consciously aims at the generation and regulation of a near future that is closely controlled by the present; and even some, perhaps the most extreme, forms of religious fundamentalism affirm a destruction of the values of the present to make way for a terrifying or welcoming unknown, drawn from the antiquarian "authenticity" of the past, which is to follow.) Politics is an address to an immediate or middle-term future which attempts to refigure the value of the past through a critique of the present. It is a contestation of how the past is read and what of the past still subsists in present activities, still surges with a virtuality that makes practice in the future able to emerge. But above all, it is about how the past and present can dissociate to bring about something new.

Feminist politics and the theories that underlie it are all, whether aware of it or not, engaged in this split or dislocated duration. An excessive product or effect of a patriarchal system in which women and femininity are necessary but do not define themselves or their world, feminism is one of the uncontrolled actualizations of patriarchy's virtual force (along with the expected patriarchal institutions and practices). It is not an inevitable effect of patriarchy, but lies latent or potential within patriarchal relations, which must, whatever the scope and extent of male privilege, integrate themselves with a nature that has bifurcated or divided the sexes such that each requires the other for reproduction and the continuity of generation, each must live with or near the other. Feminism is able to challenge, critique, struggle against patriarchy because it is spawned by it, knows it intimately, is its underside and excess. But feminist struggles of all kinds aim to produce a breach between the overwhelming weight of the patriarchal (or racist) past, its disruption in the present (which is to some extent controllable), and its overcoming in the future (which is not controllable or predictable). Feminist theory, in particular, though no less than feminist political struggles, is about the future, what is to come, what cannot be foreseen, what must be made rather than known. It is about making the categories of men and women and their relations—the realm of sexual difference—different from the ways they currently exist, giving them an open future, granting each the awe and, as Irigaray describes it, the surprise of an encounter between two beings who may begin to know their difference.

Feminist struggles remain necessary into the foreseeable future, the future that still directs itself to the human, because the relations between the sexes are a provocation that will not go away, a biological inheritance that remains intractable: biology dictates that to the degree that sexually dimorphic beings emerge, their sexual difference diverges more over time. The sexes become more different over time. The human as an inclusive, singular form becomes more attenuated as the human evolves in its two divergent directions. Feminism remains necessary to the extent that we need to continue to consider what the terms male and female, man and woman mean now and will come to mean, what their relations are and will be, in what their differences consist or could consist, and how they can come to be known. As the most direct and invested response to the question of the relations between the sexes, feminism remains necessary to the extent that sexual difference continues to be a question, a mystery, a struggle, a hostility, that is, as long as the human remains human and perhaps even beyond the human. Provoked as an excess of patriarchy—which is the system, or one of them, that attempts to regulate sexual/economic/social relations between men and women according to the interests of men—feminism is as virtual in the biology of sexual difference as patriarchy itself. Patriarchy is possible, and feminism remains necessary, to the extent that the human does not know itself and is open to transformation in various directions, according to different forces and interests, that is, as long as the human evolves.

To address the future of sexual difference differently, feminism must insist on the (biological/cultural/phenomenological) irreducibility of the question of sexual difference, that is, must regard sexual difference as a continuing question, but also must devise different ways of knowing, different models of science, different forms of intuition and representation by which such differences can be understood and made anew. Knowledges that direct themselves to differences in kind, to organic continuity and its natural articulations, to qualitative and durational "objects" and problems are quite rare in the history of Western thought, though not entirely absent. They appear in the strangest places, sometimes as marginal to their fields (as Nietzsche was), sometimes as central (as Darwin was) or both (as Bergson was). But until new ways of knowing are developed or reactivated, sexual difference, and all the mysteries of living bodies, will remain tied to calculation rather than elaborated in (uncontrolled, that is, artistic) experiment.

Like any radical politics, feminism is a gamble, for it can guarantee no outcomes and always contains the risk of loss and failure. It is an ongoing struggle, for it is the articulation of ways of living, an ongoing experiment in the attainment of maximal difference rather than the attainment of specific goals. It is an art more than a science, a mode of intuition rather than reflection, dealing with bringing into existence new social relations, distributions of force, theoretical models, concepts, and ethical values, the likes of which have never existed before. Politics is an invention, a labor of fabrication, of experimentation with the unrepeatable and the singular, that links it more to intuition, to artistic production and aesthetic discernment than to planning, policy, or the extrapolation of existing relations. Clearly, both tendencies—the joy of artistic experimentation with no predictable end and the concern with the detailed implementation of plans and principles—exist in feminist and other forms of radical race, class, and sexual politics; however, most theory and practice have been directed more to extending the present into the foreseeable future, with planning, legislation, and extrapolation, than with the risk of a leap into the unknown. Yet, new practices and the constitution of new relations imply just such a risk.

It is largely political groups and technological inventors who hope for a future that is different from the present. But these two categories operate in fundamentally different ways and with quite different objectives. Technological invention seeks out a future still recognizable in present terms, a future that promises a measurable improvement over the present's use of resources. Such invention promises us now what we can still recognize as useful, an advance on what we presently have. It is governed by utility, which, as Nietzsche understood, always implies a precedence of current needs or wants. It is, in Bergson's sense, an imminent future, one that can be extrapolated from the present by degrees, one to which we can become habituated.

Political struggle, by contrast, is directed, at its best, to a future we cannot directly recognize, a future that does not simply extend our current needs and wants but may actively transform them in ways we may not understand or control. Feminist struggle, at its best, is not directed simply to the extension to women of the privileges previously accorded to men; queer struggles are not oriented simply to the attainment of heterosexual goals and ideals. At their most critical and incisive, such struggles aim for a future in which the sexes are no longer recognizable in their present terms, in which sexuality and sexual pleasure have been redefined beyond their current divisions, categories, and activities, in which identity is seen in terms of what is to come rather than what has been, where struggles are undertaken even in the knowledge that they will not solve current problems and questions so much as redirect them in different terms. The most radical and deeply directed projects of feminist, queer, antiracist, and postcolonial struggles involve a welcoming of the unsettling of previous categories, identities, and strategies, challenging the limits of present divisions and conjunctions, and reveling in the uncontainability and unpredictability of the future.

Activists worry that, without some preparation and planning for such a future, political struggles in the present will remain directionless, unable to challenge and dismantle their objects of contestation, the dominant regimes of power, unable to provide a positive alternative to what exists now. But even with careful planning and preparation, the political alternatives to present domination are not there, simply waiting to be chosen, possible but not yet real. These alternatives, as Bergson recognized, are not alternatives, not possibilities, until they are brought into existence. The task is not so much to plan for the future, organize our resources toward it, to envision it before it comes about, for this reduces the future to the present. It is to make the future, to invent it. And this space, and time, for invention, for the creation of the new, can come about only through a dislocation of and dissociation with the present rather than simply its critique. Only if the present presents itself as fractured, cracked by the interventions of the past and the promise of the future, can the new be invented, welcomed, and affirmed.

NOTES

1 Introduction

1 This becomes increasingly more pressing as biomedical research focuses in ever greater detail on the smaller and smaller components of biological existence: hormones, proteins, cells, the gene itself. The more technologies render microscopic elements of the body knowable, the more a philosophical understanding of the place of biology in culture is required as a necessary counterbalance.

2 In *Volatile Bodies* (Grosz 1994).

3 This is a claim developed with great insight in Pheng Cheah, "Mattering" (1996), and especially in *Spectral Nationality* (2003).

4 In *Space, Time and Perversion* (Grosz 1995), where for some strange reason, and in spite of the title, the question of time is barely mentioned; and in *Architecture from the Outside* (Grosz 2001), where I explore some of the same figures (Bergson, Deleuze, Irigaray) as here, but in terms of their implications for considering space.

5 This concept of the untimely is ultimately Nietzschean, but it has been carefully elaborated by both Derrida (1995b) and Deleuze (1997).

6 Irwin C. Lieb, *Past, Present and Future* (1991) discusses in more detail than I some of the contributions of these figures regarding the reality of time.

7 See de Landa (1999).

8 While Deleuze (1983) makes clear Nietzsche's untimeliness, Derrida too is interested in the disruptions of time, in the ghostly, the haunted, that which returns unseasonably (see Derrida 1992, 1995 a, 1995b).

9 This concept of the virtual has been of central concern to a number of theorists in philosophy, cultural studies, and architectural theory for a number of years. See, for example, Kwinter (2001), de Landa (2002), Rajchman (1992, 1998, 2000),

Adamson (2002), and Shields (2003). The impact of a concept of the virtual on feminist thought is still unclear, though it seems to have clear resonances with much of Irigaray's work on the interconnectedness, the isomorphism, of individual and social relations.

10 In this context, Deleuze (1983, 1988) is of primary relevance, but so too is Deleuze (1994) and Deleuze and Guattari (1987).

1 Darwinian Matters

1 I do not want to imply that nature and matter should be too closely identified, though they are clearly linked. There is nothing material that isn't in some sense derived from or obtained through nature; even synthesized materials, those produced through technologies, are in some sense part of the natural world and its laws, principles, and constraints. Matter is both more and less than the world of nature: it is more than nature insofar as there are now materials produced that are never found as such in nature, or at least without the intervention of human practices into natural processes; it is less insofar as the entire aggregate of matter leaves out the principles regulating the natural world that enable the coexistence and interrelations of material objects with each other. Nature implies the possibility of an order, a structure or form, a systematicity that matter can have only in its own (natural or cultural) context.

2 See, for example, Dennett (1996), Oyama (2000a, 2000b), Sterelny (2001), and Godfrey-Smith (1994) for recent examples of this growing interest.

3 This is a task ably undertaken in Depew and Weber (1997).

4 Compare "If there is one thing which the *Origin* is not about, it is the origin of species," quoted in Jeff Wallace (1995, 4).

5 There are a number of remarkable coincidences between the positions of the two Darwins, and in many ways Erasmus, just as easily as Charles, could be designated as the discoverer of the theory of natural selection. However, like the works of Wallace, Lamarck, Huxley and others, there are still major differences. One quote from Erasmus Darwin's *Zoonomia* (1794) will make the influence of the senior on the junior Darwin more obvious:

> From thus meditating on the great similarity of the structures of warm-blooded animals, and at the same time of the great changes they undergo both before and after their nativity; and by considering in how minute a portion of time many of the changes of animals . . . have been produced, would it be too bold to imagine, that in the great length of time, since the earth began to exist, perhaps millions of ages before the commencement of the history of mankind, would it be too bold to imagine, that all warm-blooded animals have arisen from one living filament, which THE GREAT FIRST CAUSE endured with animosity, with the power of acquiring new parts, attended with

> new propensities, directed by irritations, sensations, volitions, and associations; and thus possessing the faculty of continuing to improve by its own inherent activity, and of delivering down those improvements by generation to its posterity, world without end! (quoted in Dyson 1997, 18–19)

In his understanding of descent and of the modification and adaptation of characteristics, Erasmus closely anticipates Charles's more fully elaborated position, though in other respects, Erasmus's claims were closer to Lamarck's position than to Charles's. For further details on the relations between Erasmus, Samuel Butler, and Charles Darwin, see Dyson (1997).

6 There are, of course, a number of other figures who closely anticipated some of Darwin's central precepts, among them Thomas Malthus (1798/1976), Samuel Butler (1887), Alfred Russel Wallace (1870, 1889), Herbert Spencer (1864), and Jean-Baptiste Lamarck (1914), whose work Darwin himself discussed in careful and knowledgeable detail.

7 See, for example, Kaufman (1993).

8 Jacques Monod (1971, 23) argues that the probability for the emergence of life was extremely low, an incalculable and unlikely accident. It was a unique, and perhaps unrepeatable event, in the full sense of the word, a singularity: "We must conclude that the emergence of life on earth was probably unpredictable before it happened. We must conclude that the existence of any particular species is a singular event, an event that occurred only once in the whole of the universe, and therefore one that is also and basically unpredictable, including what species we are, namely man."

9 Unlike Monod's claim, which affirms Darwin's, that the emergence of life was extremely improbable, Stuart Kaufman (1993, xvi) argues that the probability for the emergence of something resembling life as we know it now is extremely high, the direct result of the autocatalytic properties of proteins and nucleic acids: "We can think of the origin of life as an expected emergent property of a modestly complex mixture of catalytic polymers."

10 This is the central concern of much of the work of Chris Langton (1989), the degree to which certain informational structures—computer programs, DNA maps—depend on or function independent of any particular material constellations. It is the wager of those working in the field of artificial life that there is a substrate neutrality, a pattern to life that makes it transferable from one context, computer language, or framework to another. Others more committed to questions of embryology and biological development suggest that life is tied to the forms of corporeality in which it finds itself. Emergent patterns are not those of a life somehow secreted within an organic shell but of the body's living capacities. Both strands—conceiving of life as an informational pattern, and conceiving of life as a mode of bodily development—find their source in Darwin, though each develops one of two contrary impulses in Darwin's own writings: those seized on

by geneticists, where the program is central, and those seized on by embryologists concerned with life's corporeal elaborations.

11 It is significant, as we shall see in our discussion of Bergsonism, that Bergson, too, uses the metaphors of folding and unfolding, rolling out or rolling up, that inhere in this term.

12 For further discussion of the relation between difference of degree and differences in kind, see chapter 7.

13 Darwin's close analogy between species and languages in many respects follows that outlined by his contemporary and correspondent, Charles Lyell (1863, ch. 13).

14 This, in brief, is the argument developed by Cavalli-Sforza (2000): the movement of genetically defined populations provides a broad map of the movement of linguistic populations.

15 I have discussed some of the congruences between the mappings of natural and cultural systems in "Thinking the New: Of Futures Unthought" in Grosz (1999). See also Volk (1995) and Johnson (2001).

16 See, for example, the computer modeling of ants' collective and social activities, such as building an anthill (Drogould, Ferber, Corbara, and Fresneau 1992). For collective foraging in ants, see "Modelling Foraging Behaviour of Ant Colonies" (Fletcher, Cannings, and Blackwell 1995). For examples of the modeling of human social relations, see, for example, Gilbert and Conte (1995) and Emmeche (1994). I discuss in further detail the idea of complex emergent behavior in ant society in chapter 8.

17 For an outline of the historically intricated alignment in the development of linguistics, biology, and economics out of the fields of philology, natural history, and the study of the accumulation of wealth, see Foucault (1970, 1972).

18 It is not surprising that, in the past twenty or more years, clear convergences have been recognized among the quite disparate fields of economics, linguistics, meteorology, computing, biology, and even politics. These are not structural similarities, as an earlier generation assumed, but resemblances produced by the self-organizing, bottom-up emergence of principles with no external direction or control, no mediator, commander, or creator. Steven Johnson (2001, 18), in his recent text *Emergence*, argues that "some of the great minds of the last few centuries—Adam Smith, Friedrich Engels, Charles Darwin, Alan Turing—contributed to the unknown science of self-organization, but because the science didn't exist yet as a recognized field, their work ended up being filed on more familiar shelves . . . What features do all these systems share? In the simplest terms, they solve problems by drawing on masses of relatively stupid elements, rather than a single, intelligent 'executive branch.'" See also Coveney and Highfield (1995), Casti (1990, 1994), and Cohen and Stewart (1996).

19 For further details on the relations between Darwin and Malthus, particularly

for the significant ways Darwin alters and transforms Malthusianism, see Benton (1995, 68–94).

20 See also Smith (1982).

21 Just as stable planetary orbits are constructed out of a moment-by-moment balance between inertia and gravity, then, a free market, which producers and consumers can enter and exit at will, will produce stable patterns of production, exchange, and consumption. In this way, the economic sphere will be run by laws analogous to those of a Newtonian system (Depew and Weber 1997, 116).

22 Depew and Weber's (1997) very detailed analysis of the nearly two-hundred-year history of Darwinism provides a very plausible argument for its own growing adaptation to developments, particularly in physics. Their argument, in the briefest possible terms, is that Darwinian biology, born in the wake of Newtonianism, underwent a major revitalization under the Modern Synthesis, which blended Darwinism closely with Boltzman's reduction of thermodynamics to probable arrays of molecules, which eventually culminated in quantum mechanics, and which has continued to influence evolutionary thought as it becomes more elaborated into more developed theories of self-organization.

2 Biological Difference

1 In, for example, the writings of Lewontin (1992), Lewontin, Rose, and Kamin (1984), Gould (1989, 2002), and Oyama (2000b).

2 See, for example, the quite subtle arguments of Susan Oyama (2000a, 2000b), which suggest that, even from a genetic point of view, the gene, or even the genome as a whole, cannot be regarded as a causal agent. Genotype does not cause phenotype, any more than environment causes phenotype, for genes are both more and less than cause and biological bodies are both more and less than effects. Cells are active in selecting which genetic materials are relevant to them, as much as genes form programs that distinguish the functioning of cells; moreover, neither genes nor cells can function without the active resources and input of environments, be they internal or external to the organism.

3 Ernst Mayr (1997, 11) argues that perhaps one of the reasons that evolutionary theory was articulated in Britain rather than in the more biologically sophisticated Germany was precisely the British distrust of essentialism:

> It has long been a puzzle for the historian of biology why the key to the solution of the problem of evolution was found in England rather than on the European continent. No other country in the world has such a shining galaxy of famous biologists in the middle of the last century as the Germany of Rudolphi, Ehrenberg, Karl E. von Baer, Schleiden, Leuckart, Siebold, Koelliker, Johannes Müller, Virchow, and Leydig, and yet the solution to the

> problem of evolution was found by two English amateurs, Darwin and Wallace, neither of whom had had thorough zoological training. How can one explain this? My answer is that philosophical thinking on the continent was dominated at that time by essentialism.

4
> There is no exception to the rule that every organic being naturally increases at so high a rate, that, if not destroyed, the earth would soon be covered by the progeny of a single pair. Even slow-breeding man has doubled every twenty-five years, and at this rate, in less than a thousand years, there would literally not be standing-room for his progeny . . . The elephant is reckoned the slowest breeder of all known animals, and I have taken some pains to estimate its probably minimum rate of natural increase; it will be safest to assume that it begins breeding when thirty years old, and goes on breeding till ninety years old, bringing forth six young in the interval, and surviving till one hundred years old. If this be so, after a period of from 740 to 750 years there would be nearly nineteen million elephants alive, descended from the first pair. (OS 91–92)

5 Dobzhansky (1937/1982, 179) has argued that because the environment is always changing, partly because of the depleting activities of species, natural selection requires an openness to frequently changing contexts, and this is itself a way that natural selection regulates the species' ongoing attunement to change: "The environment does not remain constant, either in terms of geological periods or even from one year to the next. Selection and mutation rates, and hence genetic equilibria, are therefore in a state of perpetual flux. The nature of the genetic mechanism is therefore such that the composition of the species population is probably never static. A species that would remain long quiescent in the evolutionary sense is likely to be doomed to extinction."

6 While the teratological influence on mutation and genetic transformation is commonly noted, there is currently a body of research on epigenetic markers that indicates a more direct relation between the forces of natural selection, or at least, environmental effects, and the heritability of genetic variations they produce: "Over the course of evolutionary time, a variety of mechanisms, mediated by epigenetic factors, have emerged to generate new variation with the potential of 'bailing out' organisms that have become dysfunctional under conditions of stress. Selection—intracellular, cell lineage, or organismic—provides the conditions under which adaptive variants can become fixed. For many organisms that normally reproduce asexually, a switch to sexual reproduction can provide this diversity" (Keller 1998, 116).

7 Mayr (1997) concurs here with the claims made, for example, in Dobzhansky (1982) and Huxley (1942).

8 As Lestienne (1998, 50) suggests:

> The chance-brought event, the "utility of the variation," is found at the intersection of the variation's chain of causality, which has its own logic, and selection's chain of causality . . . In nature there is no direct connection between the causes of variation and selection . . . It is important to recognize that this notion of a rift is distinct from—and in fact weaker than—that of independence, which is often evoked as a foundation for chance-brought phenomena. It is not necessary that there be absolute independence between the external environment of the living being that governs selection and the internal conditions that govern the appearance of variations. It is enough that this independence act in only one direction, a bit like a valve. The variations must not result in any way from selection, but the selection may result from, and in fact does often depend on, the variations.

9 See also Hodges (1983), Boden (1996), Heims (1980), Schnelle (1994), and Herken (1994).

10 See also George Kampis (1992, 1995), one of the contemporary heirs of Bergsonism, who explores biology as an open-ended observer-inclusive system instead of an objectively quantifiable one. In many ways, Kampis can be regarded as Dennett's natural adversary.

11 By the same token, to the degree to which Wallace introduces a self-contained, nonevolving, God-given mind, Darwin's contemporary, Herbert Spencer (1872), argued for precisely the converse claim: that mind itself is the same kind of automaton as body, produced through precisely the same processes, through a relation between the organism and the world. Many argue that Spencer's work provides the earliest anticipations of John von Neumann's cellular automata. See, for example, Peter Godfrey-Smith (1994, 80–89). In any case, Bergson himself regarded Spencer as his first teacher of evolution, even though, as his work develops, it grows more critical of Spencer's claims. See Bergson, *The Creative Mind* (1992).

12 For further detail on the disjunction between the evolution of the body and the spiritual progress of the mind, see also A. R. Wallace (1910) and Wallace, "Limits of Natural Selection in Human Evolution" and "Spiritualism and Human Evolution" in Camerini (2002).

13 I discuss many of these examples in my earlier work on teratology; see Grosz (1996, 55–65).

14 This notion of the brain's development providing a generalized rather than a particular or immediate response to a given stimulus prefigures Bergson's quite ingenious understanding of the brain as the locus of a certain freedom, the freedom of eliciting unpredictable or novel behavior through a kind of disconnection from the obviousness of instinctive responses. This is developed further in chapter 8.

15 "Everyone has seen how jealous a dog is of his master's affection, if lavished on

any other creature; and I have observed the same fact with monkeys. This shews that animals not only love but also have the desire to be loved" (DM 41–42).

3 Evolution of Sex and Race

1 It makes sense, given the vast amount of time required for most species to manifest observable changes (well beyond the lifetime of any human observer) that scientists must primarily focus on relatively controlled and speeded up samples or versions of selection: bacterial, viral, and insect species, which reproduce rapidly, and mathematical and computer simulations of selection. Or else they must settle for the more descriptive, historical, and speculative project of physical anthropology:

> Darwin found a deep analogy between the way breeders utilized appropriate artificial crosses to direct natural species towards new interesting varieties, and the way natural selection was acting on live forms to promote evolution. Manmade evolution is extremely rapid when compared with the products of natural evolution. However, Darwin reasoned that gradual changes acting during long periods of time, could ultimately shape the huge number of biological species and the variety of emergent biological properties that are present now or have previously disappeared. It was only a matter of time. Evidence for natural selection is mostly indirect, i.e., it is obtained when looking in the natural history repertoire and comparing patterns of change of a given character in populations and/or species. Then it is hypothesized that natural selection is responsible for the pattern observed. There is not enough time for a human observer or even a limited set of human generations, to witness the emergence of, for instance, a new species, or the functional change from an ancestral property into a new function. (Moya, Domingo, and Holland 1995, 161)

2 The "gene's-eye" point of view elaborated by E. O Wilson, Dawkins, Dennett, and others relies on this shift, latent in Darwin, from the point of view of the organism, in *Origins*, to the point of view of the gene, or reproductive success, in *Descent.*

3 See Irigaray, "Is the Subject of Science Sexed?" (1987).

4 As Darwin makes clear: "As the male has to search for the female, he requires for this purpose organs of sense and locomotion, but if these organs are necessary for other purposes of life, as is generally the case, they will have been developed through natural selection" (DM 2: 256).

5 For example, the work of Patricia Adair Gowaty on birds is largely directed to showing how a more egalitarian approach to behavioral ecology would improve science. Her feminism largely consists in asking what the females, as well as the

males, in her study contribute to the survival of the group and the species. See Gowaty (1995) and her chapter "Darwinian Feminists and Feminist Evolutionists," in Gowaty (1997b: 1–17). See also Sayers (1982).

6 I have attempted to elaborate this point, even if briefly and perhaps inadequately, in *Volatile Bodies* (Grosz 1994).

7 At the time of the publication of *Descent*, there was an outcry from a number of concerned female readers. See, for example, Antoinette Brown Blackwell's (1875, 11) position in *The Sexes throughout Nature*, where she claims that Darwin gave "undue prominence to such as have evolved in the male line." For more recent examples, see Morgan (1992), Fausto-Sterling, Gowaty, and Zuk (1997), and Gowaty (1997 b).

8 This musicality is precisely what Julia Kristeva (1984) understands as the "semiotic," corporeal component necessary for all languages, which imparts the material elements or components of representation.

9 Quite remarkably, as we will see, Weismann's (1893) distinction between germ cells and somatic cells closely anticipates the contemporary distinction between DNA and proteins.

10 See, for example, Wilson (1980, ch. 5).

11 The horrifying exception, of course, is genocide, the self-conscious attempt by some races to "artificially select" the demise of their competitors and to elevate their own prosperity and chances of survival.

12 There are some exceptions. Sickle cell anemia in African communities, while potentially deadly and commonly manifest as an inability to digest milk, also has the natural selective benefit of providing immunity to malaria. This helps explain why many white colonists succumbed readily to the deadly effects of malaria. This, however, is a rare case.

13 Cavalli-Sforza (2000), though careful in many respects, falls prey to the reduction of sexual to natural selection. Although he doesn't attribute environmental factors as the sole elements producing racial differences, nonetheless he does functionalize racial characteristics to such an extent that they seem more effects of natural than sexual selection: "The size and shape of the body are adapted to temperature and humidity. In hot and humid climates, like topical forests, it is advantageous to be short since there is greater surface area for the evaporation of sweat compared to the body's volume. A smaller body also uses less energy and produces less heat. Frizzy hair allows sweat to remain on the scalp longer and results in greater cooling. With these adaptations, the risk of overheating in tropical climates is diminished. Populations living in tropical forests are generally short, Pygmies being the extreme example. The face and the body of Mongols, on the other hand, results from adaptations to the bitter cold of Siberia" (11).

14 His relative openness to racial variations and practices sharply differentiates his

position from that of Galton, whose position is more typical, understood in the context of Eurocentrism and British colonialism:

> In his young manhood, Galton had done some geographic explorations in Africa in the fashion of Charles Darwin's travels, carrying with him not only Humboldt but the *Voyage of the Beagle* as well. On his African adventures, he had become convinced of the inherent and unamendable inferiority of the non-white races. Unlike Darwin, he did not content himself with long-term visions of human progress or insist that native populations and cultures must be free enough to develop intellectually and morally. On the contrary, he worried that the higher birthrates of the inferior races and the long-term effects of miscegenation would dissipate the superiority of the white Europeans with whom they were now in contact. These were the worries of an imperialist. They contrast vividly with Darwin's conviction that, even though the races are unequal, sexual selection . . . and the cultural transmission of traits that favor the development of some moral sensitivities, would make it possible for all to progress slowly upward. Darwin was a child of the great age of democratic revolutions. Galton was a son of the age of imperialism. (Depew and Weber 1997, 197)

15 Dennett (1996, 56) makes clear the fallacy of assuming that evolution was destined to produce *us* in precisely the form we occupy now, as the pinnacle of creation: "We can now expose perhaps the most common misunderstanding of Darwinism: the idea that Darwin showed that evolution by natural selection is a procedure *for* producing Us. Ever since Darwin proposed his theory, people have often misguidedly tried to interpret it as showing that we are the destination, the goal, the point of all that winnowing and competition, and our arrival on the scene was guaranteed . . . Evolution can be an algorithm, and evolution can have produced us by an algorithmic process without its being true that evolution is an algorithm for producing us."

16 As Irigaray (1996, 47) claims, in *I Love to You: Sketch of a Possible Felicity in History*:

> Without doubt, the most appropriate content for the universal is sexual difference . . . Sexual difference is an immediate natural given and it is a real and irreducible component of the universal. The whole of human kind is composed of women and men and of nothing else. The problem of race is, in fact, a secondary problem—except from a geographical point of view . . . and the same goes for other cultural diversities—religious, economic, and political ones.
>
> Sexual difference probably represents the most universal question we can address. Our era is faced with the task of dealing with this issue, because, across the whole world, there are, there are only, men and women.

4 Nietzsche's Darwin

1 These varying time scales, noted among many evolutionists, do not, however, mean a variety of times. Rather, they imply that different modes and organizations of materiality each have their own temporal regimes and rhythms, which cannot always be coordinated together. The movement of geological forces, for example, is exceedingly slow compared to the movement of organisms; equally, the movement or temporality of ideas and cultural institutions is rapid compared to genetic transformations. See, for example, Murray Gell-Mann, "Complex Adaptive Systems" (1994, 23), who discusses the different scales of evolutionary movement in the emergence of language: "One is the evolution over hundreds of millions of years, of the biological capacity to use languages of the modern type. Another is the evolution of those languages themselves, over thousands or tens of thousands of years. Yet another is the learning of a native language by a child." See also de Landa (1997, 1999).

2 In this and subsequent chapters on Nietzsche, I follow Nietzschean scholarship in providing title, book, and section (#) citations, relevant for all editions of his work, rather than page numbers, at least in those texts that use them. In the others, I use common citation practices.

3 Ansell Pearson (1997b, 90) argues that although Nietzsche misunderstands Darwin, he nevertheless provides an incisive and accurate critique of social Darwinism: "A fundamental aspect of the revaluation of values conducted in the genealogy or morality will be a revaluation of 'Darwinian' values. This revaluation, however . . . is not without major problems since it raises the complex issue of unwarranted anthropomorphizations of nature and corresponding reifications of natural and technical life." He goes on to carefully outline the centrality that a number of German Darwinists, notably Wilhelm Roux and Carl von Nageli (embryologists and developmentalists), have in the development of his understanding of genealogy as well as his conception of bodily organs.

4 For further detail on the idea of the will to power as a multiplicity of wills that seek to command and those that seek to obey, see Alphonso Lingis (1985, 37–63).

5 History and the Untimely

1 Deleuze (1983, 50) elaborates that the will to power is both the victorious force and the synthesized field of all the forces in their differentiation: "We must remember that every force has an essential relation to other forces, that the essence of force is its quantitative difference from other forces, and that this difference is expressed as the force's quality. Now, the difference in quantity . . . necessarily reflects a differential element of related forces—which is also the genetic element of the qualities of these forces. This is what the will to power is;

the genealogical element of force, both differential and genetic. *The will to power is the element from which derive both the quantitative difference of related forces and the quality that devolves into each force in this relation.*"

2 "The will to power cannot be separated from force without falling into metaphysical abstraction. But to confuse force and will is even more risky. Force is no longer understood as force and one falls back into mechanism—forgetting the difference between forces which constitutes their being and remaining ignorant of the element from which their reciprocal genesis derives. Force is what can, will to power is what wills" (Deleuze 1983, 50).

3 As Deleuze (1983, 51) explains: "The will to power is both the genetic element of force and the principle of synthesis of forces."

6 The Eternal Return and the Overman

1 The principle of the conservation of energy—that energy is never produced or destroyed but transferred from one form to another (mechanical, chemical, or electrical)—though a truism since the ancient Greeks, was explicitly formulated in the early nineteenth century by Mayer, and then further elaborated by Joule and Helmholtz. For further details on the nineteenth-century developments of the first and second laws (elaboration of the second law, also known as the principle of entropy, was required for a full understanding of the first), see Swenson (1997). Nietzsche was fascinated with the implications of these two principles, which form the underlying grid that the eternal return embroiders and complicates.

2 Danto (1968, 203–204) claims that although Nietzsche barely published much material on the eternal return, the *Nachlass* contains a sketch for a book provisionally entitled *The Eternal Return: A Prophecy.* This brief proposal is published in *The Will to Power* (#1057). Nietzsche had such faith in the centrality of this idea to all of his writings that at one point, at the height of his powers in the 1880s, he seriously considered returning to university to study the natural sciences, which he believed would provide support for his doctrine. As it turns out, he would have had to wait over a century to find cosmological and biological support for his account!

3 Around the same time as he was writing *The Gay Science*, Nietzsche pondered the eternal return as the supreme thought, but also as the supreme *feeling*, an intensification of both intellect and affect. Unlike many of Nietzsche's other commentators, Klossowksi emphasizes the affective dimension of Nietzsche's conjecture: "My doctrine teaches: live in such a way that you must desire to live again, this is your duty—you will live again in any case. He to whom effort procures the loftiest feeling, let him make the effort; he to whom repose brings the loftiest feeling, let him make the rest; he to whom the act of joining, of following and of obeying procures the loftiest feeling, let him obey. Providing

that he becomes aware of what procures the loftiest feeling and that he draws back before nothing. Eternity depends on it" (Nietzsche, quoted in Klossowski 1985, 110).

4 Deleuze (1994, 41) explains this entwinement of chance and necessity, difference and repetition: "The eternal return does not bring back 'the same,' but returning constitutes the only Same of that which becomes. Returning is the becoming-identical of becoming itself. Returning is thus the only identity, but identity as a secondary power; the identity of difference, the identical which belongs to the different, or turns around the different. Such an identity, produced by difference, is determined as 'repetition.' Repetition in the eternal return, therefore, consists in conceiving the same on the basis of the different."

5 Massumi (1992, 167–168) argues that this model of the eternal return has close resonances with that model of the genesis of the universe proposed by Prigogine and Stengers, and finds surprising confirmation in contemporary cosmology:

> They [Prigogine and Stengers] theorize that the virtual is inherently unstable because it is composed of different particles that are in constant flux, but in ways that do not harmonize. In the absence of matter, at maximum entropy, the turbulence of the virtual is amplified to the point of an explosive contraction releasing an unimaginable amount of pure energy. The energy is unstable as the void, and immediately dilates, creating matter. A universe is born . . . The presence of matter muffles the turbulence by giving it an outlet, by providing a dimension rigid enough to limit it but flexible enough to absorb it. What we get in the form of "chance" and indeterminacy is overflow from the actual's absorption of the virtual. After the initial contraction-dilation, the material universe goes on dilating slowly until its future is consumed by its past and it disappears into maximum entropy. Then it all starts over again . . . This amounts to a scientifically derived version of Nietzsche's theory of the eternal return of difference that is very close to Deleuze's philosophical version.

6 Indeed, this is the political and conceptual heart of Deleuze's reading of Nietzsche in *Nietzsche and Philosophy* (1983, 71): that Nietzsche is the theorist of the active becoming of the eternal return, that the eternal return is the return of difference rather than the repetition of the same, that what returns cannot be the reactive, the infirmed, the half-willed but only active will, active force, which, by affirming itself, also wills its own recurrence: "However far they go, however deep the becoming-reactive of forces, reactive forces will not return. The small, petty, reactive man will not return."

7 It is not some one thing that returns, but rather, returning itself is the "one thing which is affirmed of diversity or multiplicity. In other words, identity in the eternal return does not describe the nature of that which returns but, on the contrary, the fact of returning for that which differs. This is why the eternal

return must be thought of as a synthesis; a synthesis of time and its dimensions, a synthesis of diversity and its reproduction, a synthesis of becoming and the being which is affirmed in becoming, a synthesis of double affirmation" (Deleuze 1983, 48).

David Wood (2001, 32) suggests, though perhaps not without some conceptual violence, that it is at this point that Derrida's reading of Nietzsche converges with Deleuze's: "For both Deleuze and Derrida, the key underlying idea is that identity is not a fixed point one needs to presuppose for differences to be possible; matters are rather the other way round. And the possibility that a thing can appear again and again at different times is what *gives* it an identity; it is not dependent on having a prior atemporal identity. Time, then, is not only *constitutive* of identity, rather than a mere medium in which things unfold, but is itself constituted by its role in supporting identities and differences."

Wood conflates Deleuze's and Nietzsche's positions too closely with his own reading of Derrida's. Neither Nietzsche nor Deleuze, whatever their major differences, ties the concept of time to "supporting" either identities or differences (differences here are reduced to nonidentities, identities by another name). Time is itself pure difference. It would seem that the status of repetition, which Deleuze, following Nietzsche, links directly with the force of difference, is different from the Derridean concept of iteration, the repetition of the "same" in an always different, ultimately indeterminable context.

8 As Deleuze (1983, 47) glosses: "If becoming becomes something, why has it not finished becoming long ago? If the world had a goal, it must have been reached. If there were for it some unintended final state, this also must have been reached. If it were in any way capable of a pausing and becoming fixed, of 'being,' if in the whole course of its becoming it possessed even for a moment this capability of 'being,' then all becoming would long since have come to an end, along with all thinking, all 'spirit.' The fact of 'spirit' as a form of becoming proves that the world has no goal, no final state, and is incapable of being."

9 There are striking parallels between Linde's (1994) hypothesis of inflationary universes and Nietzsche's understanding of the eternal return. The Big Bang can be regarded as a kind of inflationary bubble of energy compressed into dense matter. This bubble cannot be regarded as the one and only eruption of matter into existence, for it is one in a long chain of bubbles, each eventually exploding to form a universe of heat energy, condensing to form matter, and winding down to a level of low energy which then becomes bubble-like and begins again. One inflationary bubble would trigger another, which would trigger many more, to infinity. Each bubble is in effect a new Big Bang and a new universe ad infinitum. For further details on this conception of cosmological becoming, see Linde (1994) and Guth and Lightman (1998).

10 Peter Berkowitz (1995, 177) is one who remains highly critical of the attempts to justify the physical doctrine as central to the moral doctrine on the part of

Nietzsche's commentators, among them, Nehemas (1985), Strong (1988), and Lampert (1986). Instead, he argues that the failure of the doctrine as both physical description and ethical imperative marks the failure of Zarathustra's goal of god-like mastery of eternity. Berkowitz, in other words, affirms Heidegger's (1979, 18–24; 1984, 98–105) provocation that the eternal return is Zarathustra's revenge on transience, on mortality (Zarathustra regards himself as beyond the spirit of revenge).

11 I teach you the overman. Man is something that shall be overcome. What have you done to overcome him?

> All beings so far have created something beyond themselves; and do you want to be the ebb of this great flood and even go back to the beasts rather than overcome man? What is the ape to man? A laughing-stock or a painful embarrassment. And man shall be just that for the overman: a laughingstock or a painful embarrassment. You have made your way from worm to man, and much in you is still worm. Once you were apes, and even now, too, man is more ape than any ape.
>
> Whoever is wisest among you is also a mere conflict and cross between plant and ghost. But do I bid you become ghosts or plants?
>
> Behold, I teach you the overman. The overman is the meaning of the earth. Yet you will say: the overman *shall be* the meaning of the earth! (*Z*, prologue, #3)

12 It is necessary to make explicit, in a way that Nietzsche cannot afford to, that the overman is no representative of man after man's demise, but is the open possibility of what is highest in man rising above man himself. Though Nietzsche did not ask this question, following Irigaray's reading of his works, especially *Zarathustra*, in *Marine Lover* (1991), it is necessary to inquire about whether, for this overman, this overhuman, there is also a comparative overwoman: that is, to ask how this future of mankind remains marked by its own irreducible differences. This question, the future of the differentials dividing man, is one, it seems, that Darwin regarded in more open terms than Nietzsche! This overcoming of man is not a reduction of man but a complexification. If man is developed, if mankind is developed, there must be a magnification and expansion of sexual difference itself.

13 Erich Heller (1988, 183) argues that the eternal return is the test that distinguishes the exceptional individual from the herd: the herd is destroyed by this idea that only the exceptional one can tolerate and welcome.

14 As Wood (2001, 19–20) suggests: "Eternal recurrence seems merely to be a very powerful *modification or extension of seriality.* But in fact it puts into question the assumed self-contained status of its units—'nows'—and their fundamental successivity. For as well as being located on a horizontal axis of succession, each 'unit' also appears to be a member of a vertical series of repetitions."

7 Bergsonian Difference

1 Deleuze (1988, 13) claims that intuition is "one of the most fully developed methods in philosophy. It has strict rules, constituting that which Bergson calls 'precision' in philosophy." Elsewhere, Deleuze describes intuition as "the *jouissance* of difference" (1999, 43).

2 Among these detractors, the most vicious was undoubtedly Bertrand Russell (1912, 1914, 1957), a theorist not known for his subtlety of reading practices, who was strongly supported in style and criticism by Julian Huxley (1926). Russell challenged a number of Bergson's most central precepts, particularly his understanding of the interpenetrating nature of continuity (for a detailed discussion of Russell's mathematical understanding of continuity, the centrality of discontinuity in his work, and his strained relation to Bergson, see Pearson [2002, 24–28] and Adamson [2002, 27–35]).On the other hand, most of his other critics, in spite of their various objections or uneasiness, retain something of a sympathy for his claims and an acknowledgment of the profundity of his insights. See, for example, his contemporary critics, such as Stewart (1911) and Alexander (1957), or those who came later, such as Bachelard (2000), who significantly share Russell's suspicion of Bergson's commitment to the concept of continuity, without the same sense of scorn; as well as Merleau-Ponty, "Bergson in the Making" in *Signs* (1964) and "Bergson" in *In Praise of Philosophy* (1970). See also Levinas (1987), Lacey (1989), and Mullarkey (1999a, 1999b).

3 For Bergson's ongoing relevance to the natural sciences, see, for example, Gunter (1969, 1971), Čapek, (1976, 1991), Kampis (1991, 1993, 1995), Kampis and Weibel (1992), and Adamson (1999, 2002). For his ongoing relevance to philosophy, see Deleuze (1988), Boundas (1993, 1996), Pearson (1997a, 1999, 2002), Mullarkey (1999a, 1999b), Kwinter (1992, 1993, 2001), de Landa (2002), and Massumi (2002).

4 It is not quite precise to say that Bergson does not have an ethical project in mind. Clearly, many of his concerns, particularly in his later works, are directed to the question of ethics. His underdiscussed 1932 (1986) text, *The Two Sources on Morality and Religion*, is the focus of his understanding of ethics and religion. I have left out discussion of this text here only to concentrate more directly on his explicitly Darwinian writings and to leave the question of ethics, in both Nietzsche and Bergson, largely in the shadows, though not completely out of sight.

5 Numbers are themselves a significant elaboration or expression of unity (or identity) and multiplicity (or difference). A number is a synthesis of a particular kind of unity and multiplicity. It is a collection of one larger or many smaller individual units which must be construed as identical in order for their multitude to be discerned. Identical units form a collection, but in this case, they must remain distinct from each other, retaining their identity even as they are grouped together into a new unity. This is possible only in abstract (or real) space, which nevertheless allows us to separate identical entities in terms of their location

rather than in terms of their qualities. Number neutralizes qualitative features to mark out quantitative degrees or resemblances. See Bergson (1959, *Time and Free Will*, 75–85). See also Turetzky (1998, 194–196) and Adamson (2002, 40–44).

6 For further elaboration of noncomparable differences, of difference without identity, see Irigaray (1985), Whitford (1991), and Braidotti (1994, 2002).

7 Bergson prefers the term "duration" (*durée*) rather than "time," which already marks or registers the encroachment of spatialized models onto the movement of temporality. Time is the spatialized, measurable counterpart of space, the other of space; duration is altogether different, a term inassimilable to any spatialized representation.

8 This occupies Bergson for the whole first chapter of *Time and Free Will*, especially 56 and after.

9 Gaston Bachelard is increasingly hostile to the continuist model to which Bergson adheres, and many of his most stringent criticisms are directed to the necessity and desirability of inserting stops, gaps, and breaks into duration, to render duration dialectical, which for him is to render it a site of oppositional strategies and scientific contestations. But in doing so, he returns to Bergson's insight that such continuity is given in consciousness, given in life. In attempting to limit and contain the psychical and the experiential outside the scientific, Bachelard simply continues the tradition Bergson criticizes without recognizing that Bergson's criticisms can be equally leveled at his own claims. As Bachelard (2000, 28–29) facetiously affirms:

> Since any critique is clarified by its end-point, let us say straightaway that of Bergsonism we accept everything but continuity. Indeed to be even more precise, let is say that from our point of view also continuity—or continuities—can be presumed as characteristics of the psyche, characteristics that cannot however be regarded as complete, solid, or constant. They have to be constructed. They have to be maintained. Consequently we do not in the end see the continuity of duration as an immediate datum but as a problem. We wish therefore to develop a discontinuous Bergsonism, showing the need to arithmetise Bergsonian duration so as to give it more fluidity, more numbers, and also more accuracy in the correspondence the phenomena of thought exhibit between themselves and the quantum characteristics of reality.

In this context, Ann Game (1995) provides further elaboration of Bachelard's relation to Bergson.

10 Intensity, like the concept of time itself, is always already the marking of the extensive onto the qualitative, a hybrid product: "Representative sensation, looked at in itself, is pure quality: but, seen through the medium of extensity, this quality becomes in a certain sense quantity, and is called intensity . . . Now, let us notice that when we speak of *time* [in the sense of duration reduced to spatializa-

tion], we generally think of a homogeneous medium in which our conscious states are ranged alongside one another as in space, so as to form a discrete multiplicity. Would not time, thus understood, be to the multiplicity of our psychic states what intensity is to certain of them—a sign, a symbol, absolutely distinct from true duration?" (TFW 90).

11 Deleuze (1999, 49–50) stresses that Bergson's work cannot be construed as binarized, for the continuous term always includes, encloses, the discontinuous:

> Duration is only one of the two tendencies, one of the two halves; but if it is true that in all its being it differs from itself, does it not contain the secret of the other half? How would it still leave outside of itself *that from which* it differs, the other tendency? If duration differs from itself, that from which it differs, is still duration, in a certain way . . . the mixture decomposes itself into two tendencies, one of which is the indivisible, but the indivisible differentiates itself into two tendencies, the other of which is the principle of the divisible. Space is decomposed into matter and duration, but duration differentiates itself into contraction and relaxation, relaxation being the principle of matter. Organic form is decomposed into matter and the *élan vital* differentiates itself into instinct and intelligence, intelligence being the principle for the transformation of matter into space.

12 The doctrine of images has had tremendous impact on Deleuze's understanding of cinema, and ironically, given what Bergson perceives as the limit of cinematographic representation (that everything on celluloid is given in advance: film represents the semblance of duration but not its own quality of open unpredictability), it has been increasingly influential in contemporary cinema studies. See, for example, Rodowick (1997), Buchanan (1997), Douglass (1992, 1998), and Bogue (2003).

13 Merleau-Ponty acknowledges the powerful influence of Bergson's understanding of the positioning of the body as an object in the world in both his earlier writings (*The Structure of Behavior* [1983], *The Phenomenology of Perception* [1962], *The Primacy of Perception* [1963]) as well as in his final texts (*The Visible and the Invisible* [1968]). What Merleau-Ponty and Bergson share is an ontology of becoming, an ontology in which consciousness and life, respectively, do not find themselves in a world but make themselves subjects, and make the world into things, objects, entities through their activity, their engagement, their labor. Active becoming is emergent. It elaborates itself from and on a field of active forces as their contingent frame. I have discussed the complex and ambivalent relations between Bergson and Merleau-Ponty in "Merleau-Ponty, Bergson, and the Question of Ontology," a chapter in *Time Travels: Feminism, Nature, Power* (2005).

14 It is at this point that the differences between Bergson and Nietzsche become explicit. For Bergson, nothing returns, nothing repeats, for time is always in-

creasing elaboration, and repetition, at least formal or nominal repetition, contains within itself an index of its prior iterations. For Bergson, formal repetition constitutes the possibility of their being counted, indeed, is the condition of number itself.

15 "Here is a system of images which I term my perception of the universe, and which may be entirely altered by a very slight change in a certain privileged image—*my body.* This image occupies the center; by it all the others are conditioned; at each of its movements everything changes, as though by a turn of the kaleidoscope. Here on the other hand, are the same images, but referred each one to itself, influencing each other no doubt, but in such a manner that the effect is always in proportion to the cause: this is what I term *the universe.* The question is: how can these two systems coexist, and why are the same images relatively invariable in the universe and infinitely variable in perception?" (*MM* 25).

16 "How is it that the same images can belong at the same time to two different systems: one in which each image varies for itself and in the well-defined measure that it is patient of all real action of surrounding images; and another in which all images change for a single image and in the varying measure that they reflect the eventual action of this privileged image?" (*MM* 25).

17 This, in the most simplified form, is the basic argument of *Time and Free Will.*

18 As Bergson suggests, to recognize something we must "assume the consciousness of a well-regulated motor accompaniment, of an organized motor reaction . . . At the basis of recognition there would thus be a phenomenon of motor order . . . If, then, every perception has its organized motor accompaniment, the ordinary feeling of recognition has its root in the consciousness of this organization" (*MM* 93–94).

19 Bergson suggests that it makes perfect sense for a dog to be bound up with habit-memory, which has potentially immediate or direct effects on its present and imminent actions. The dog may also have access to memory-images, "pure memory," but it makes no sense to see the animal detaching from its immediate present to make creative use of them:

> When a dog welcomes his master, barking and wagging his tail, he certainly recognizes him; but does this recognition imply the evocation of a past image and the comparison of that image with the present perception? Does it not rather consist in the animal's consciousness of a certain special attitude adopted by his body, an attitude which has been gradually built up by his familiar relations with his master, and which the mere perception of his master now calls forth in him mechanically? We must not go too far; even in the animal it is possible that vague images of the past overflow into the present perception; we can even conceive that its entire past is virtually indicated in its consciousness; but this past does not interest the animal

> enough to detach it from the fascinating present, and its recognition must be rather lived than thought. (*MM* 82)

20 See *MM* 112–120 and *CM* 86 on the question of aphasia.

21 Bergson cites some significant collateral symptoms of aphasia and apraxia that problematize their classification as disorders of memory. Not only are aphasics unable to recall or sometimes even recognize familiar words, which have led us, falsely, to believe that certain memories have been destroyed due to cerebral lesions; they also experience a significant loss in the sense of direction and in the ability to draw. These tend to indicate that the loss is not circumscribed to particular memories but to a more general bodily or corporeal schematization, later described by Paul Schilder (1978) as the corporeal schema, which links bodily comportment to visual perception: "All those who treat psychic blindness [i.e., apraxia] have been struck by this peculiarity. Lissauer's patient had completely lost the faculty of finding his way about his own house . . . while blind men soon learn to find their way, the victim of psychic blindness fails, even after months of practice, to find his way about his own room. But is not this faculty of orientation the same thing as the faculty of coordinating the movements of the body with the visual impression, and of mechanically prolonging perceptions in useful reactions?" (*MM* 96–97).

22 For further elaboration, see Cariou (1999, 108–109).

23 "While external perception provokes on our part movements which retrace its main lines, our memory directs upon the perception received the memory-images which resemble it and which are already sketched out by the movements themselves. Memory thus creates anew the present perception by reflecting upon it either its own image or some other memory-image of the same kind . . . And the operation may go on indefinitely—memory strengthening and enriching perception, which, in its turn becoming wider, draws into itself a growing number of complementary recollections" (*MM* 101).

24 "In other words, personal recollections, exactly localized, the series of which represents the course of our past existence, make up, all together, the last and largest enclosure of our memory. Essentially fugitive, they become materialized only by chance . . . But this outermost envelope contracts and repeats itself in inner and concentric circles, which in their narrower range enclose the same recollections grown smaller, more and more removed from their personal and original form, and more and more capable, from their lack of distinguishing features . . . There comes a moment when the recollection thus brought down is capable of blending so well with the present perception that we cannot say where perception ends or memory begins" (*MM* 106).

25 Bergson's gloss on figure 1 is as follows:

> Of these different circles of memory . . . the smallest, A, is the nearest to immediate perception. It contains only the object O, with the afterimage

which comes back and overlies it. Behind it, the larger and larger circles B, C, D correspond to growing efforts at intellectual expansion. It is the whole of memory . . . that passes over into each of these circuits, since memory is always present; but that memory, capable, by reason of its elasticity, of expanding more and more, reflects upon the object a growing number of suggested images—sometimes the details of the object itself, sometimes concomitant details which may throw light upon it. Thus, after having rebuilt the object perceived, as an independent whole, we reassemble, together with it, the more and more distant conditions with which it forms one system . . . We call B′, C′, D′, these causes of growing depth, situated behind the object and virtually given with the object itself. (*MM* 104–105)

26 In Derridean terms, and in spite of Derrida's own criticisms in "*Ousia and Grammé*" (1982), the Bergsonian present does not succumb to a philosophy of presence: his present is never self-identical, never able to be definitely separated from the past that contextualizes it and the immediate future that it functions to anticipate. It is a term that blends into its others, that never denies its dependence on the past (not the past as cause, that is, as present, but the past as virtual, as a multifaceted crystal whose virtual faces are illuminated by different presents).

27 As Deleuze (1988, 56) argues: "Strictly speaking, the psychological is the present. Only the present is 'psychological'; but the past is pure ontology; pure recollection has only ontological significance."

28 For Deleuze (1988, 56–57), this concept of "the past in general" is the ontological element of passage, of change, that enables both past and present to be distinguished: "In the same way that we do not perceive things in ourselves, but at the place where they are, we only grasp the past at the place where it is in itself, and not in ourselves, in our present. There is therefore a 'past in general' that is not the past of a particular present but is like an ontological element, a past that is eternal and for all time. It is the past in general that makes possible all pasts. According to Bergson, we first put ourselves back into the past in general: He describes in this way the *leap into ontology*. We really leap into being, into being-in-itself, into the being in itself of the past. It is a case of leaving psychology altogether. It is a case of an immemorial or ontological Memory."

29 "So, if we are right, the *hearer* places himself at once in the midst of the corresponding ideas, and then develops them into acoustic memories which go out to overlie the crude sounds perceived, while fitting themselves into a motor diagram. To follow an arithmetical addition is to do it over again for ourselves. To understand another's words is, in like manner, to reconstruct intelligently, starting from the ideas, the continuity of sound, which the ear perceives" (*MM* 116–117).

30 It is worth comparing Bergson's early paper "On Dreams" (1901) in *Mind-Energy* (1921) to Freud's *Interpretation of Dreams* (1905), which it anticipates in quite

remarkable degree. Bergson there explains that dreams occur because the inhibitions of memory that action or consciousness effect are temporarily cast aside in sleep, enabling otherwise "repressed memories" (his term) to surface in consciousness:

> Our memories, at a given moment, form one solidary whole, a pyramid whose point coincides with our present,—with a present moving ceaselessly and plunging into the future. But, behind the memories which crowd in upon our present occupation and are revealed by means of it, there are others, thousands on thousands of others, below and beneath the scene illuminated by consciousness. Yes, I believe our past life is there, preserved even to the minutest details; nothing is forgotten; all we have perceived, thought, willed, from the first awakening of our consciousness, persists indefinitely. But the memories which are preserved in these obscure depths are for us in the state of invisible phantoms. They aspire, perhaps, to the light: they do not even try to rise to it; they know it is impossible and that I, a living being, have something else to do than occupy myself with them. But suppose that, at a given moment, I become *disinterested* in the present situation . . . suppose, in other words, I fall asleep: then these repressed memories . . . raise the trapdoor which held them back below the floor of consciousness, begin to stir. They rise and spread abroad and perform in the night of the unconscious a wild phantasmagoric dance. They rush together to the door which has been left ajar. (ME 94–95)

Bergson was fascinated with the ways sleep and dreaming utilize a nonpractical form of memory. His arguments are more fully developed in *The World of Dreams* (1958).

31 "This time-image extends naturally into a language-image, and a thought-image. What the past is to time, sense is to language and idea to thought. Sense as past of language is the form of its pre-existence, that which we place ourselves in at once in order to understand images of sentences, to distinguish the images of words and even phonemes that we hear. It is therefore organized in coexisting circles, sheets or regions, between which we choose according to actual auditory signs which are grasped in a confused way. Similarly, we place ourselves initially in the idea; we jump into one of its circles in order to form images which correspond to the actual quest" (Deleuze 1989, 99–100).

32 This is Bergson's argument in *Duration and Simultaneity* (1922/1965).

33 In Deleuze's (1988, 61–62) reading, Bergson systematically develops a series of paradoxes regarding the past and present which run counter to a more common, everyday understanding. They are: "(1) we place ourselves at one, in a leap, in the ontological element of the past (paradox of the leap); (2) there is a difference in kind between the present and the past (paradox of Being); (3) the past does not follow the present that it has been, but coexists with it (paradox of coexistence);

(4) what coexists with each present is the whole of the past, integrally, on various levels of contraction and relaxation (*détente*) (paradox of psychic repetition)."

These Bergsonian paradoxes, which are only paradoxical if duration is represented on the model of space, are all, Deleuze claims, a critique of more ordinary theories of memory, whose propositions include: "(1) we can reconstitute the past with the present; (2) we pass gradually from the one to the other; (3) that they are distinguished by a before and an after; and (4) that the work of the mind is carried out by the addition of elements (rather than by changes of level, genuine jumps, the reworking of systems)."

8 The Philosophy of Life

1 In a later paper, "Soul and Body," published in *Mind-Energy* (1921), Bergson elaborates this metaphor of rolling and unrolling: "In time, for the body is matter, matter is in the present, and, if it be true that the past leaves there traces of itself, they are not traces of the past except for a consciousness perceiving them and interpreting what it perceives by the light of what it remembers. This consciousness retains the past, enrolls what time unrolls, and with it prepares a future which it will itself help to create" (30).

2 See for example, Kwinter (1993, 2001), Rajchman (1998, 2000), Massumi (2002), and Shields (2003).

3 It is Bergson's argument that disorder, absence, and the mirror image constitute more rather than less order, presence, and reality that seems to most distress Bachelard, who makes it the center of his objections in *The Dialectic of Duration* (2000).

4 Compare Bergson: "One might as well claim that the man in flesh and blood comes from the materialization of his image seen in the mirror, because in that real is everything to be found in this virtual image with, in addition, the solidity which makes it possible to touch it. But the truth is that more is needed here to obtain the virtual than is necessary for the real, more of the image of the man than for the man himself, for the image of the man will not be portrayed if the man is not first produced, and in addition one has to have the mirror" (CM 102).

5 This is the discourse, articulated by Gilbert Simondon, that has been so influential in much of Deleuze's writings, especially in *A Thousand Plateaus* (Deleuze and Guattari 1987), regarding the processes of individuation in terms of a series of states of metastable equilibrium, and thus irreducibly in terms of processes of becoming. Simondon may have succeeded in going a step further than Bergson in thinking the implications of movement as the internal condition of individuation or being itself. Individuation is a series of processes of radical excentering and self-exceeding (even at the nonorganic level of the crystal):

> The concept of being that I put forward, then, is the following: a being does not possess a unity in its identity, which is that of the stable state within

which no transformation is possible; rather, a being has a *transductive unity*, that is, it can pass out of phase with itself, it can—in any arena—break its own bounds in relation to its *center.* What one assumes to be a *relation* or a *duality of principles* is in fact the unfolding of the being, which is more than a unity and more than an identity; becoming is a dimension of the being, not something that happens to it following a succession of events that affect a being already and originally given and substantial. Individuation must be grasped as the becoming of the being and not as a model of the being which would exhaust his signification . . . Instead of presupposing the existence of substances in order to account for individuation, I intend, on the contrary, to take the different regimes of individuation as providing the foundation for different domains such as matter, life, mind and society. (Simondon 1993, 311–312)

6 If there is a movement of differentiation from a virtual unity, the unity of the past as a whole contracted in different degrees in the process of actualization, Deleuze (1988, 29; unlike Bergson at this point) suggests that there is also a complementary movement from the actual multiplicity to the virtual underlying it: "The real is not only that which is cut out according to natural articulations or differences in kind; it is also that which intersects again along paths converging toward the same ideal or virtual point." This point of convergence, reconfiguring the movements of divergence and differentiation that made a process of actualization of the virtual occur, is of course the point at which memory is reinserted into perception, the point at which the actual object (re)meets its virtual counterpart.

Deleuze wants to make this moment of convergence central to his understanding of what he calls the "crystal structure" of the time image in cinema. The crystal image is the coalescence of an actual image with "its" virtual image, a two-sided image, with one face in perception, and thus directed toward the present, the actual, while the other is steeped in recollection, in the past, the virtual: "What constitutes the crystal-image is the most fundamental operation of time: since the past is constituted not after the present that it was but at the same time, time has to split itself in two at each moment as present and past which differ from each other in nature, or, what amounts to the same thing, it has to split the present in two heterogeneous directions, one of which is launched towards the future while the other falls into the past . . . In fact the crystal constantly exchanges the two distinct images which constitute the actual image of the present which passes and the virtual image of the past which is preserved: distinct and yet indiscernible, and all the more indiscernible because distinct, because we do not know which is one and which is the other" (Deleuze 1989, 81).

The crystal image, a central mechanism in modernist cinema, is of the very essence of time: it is duration itself which splits every image into a duality of actual and virtual. It is this duality of the image, the fact that as it is created each image is placed simultaneously in time (duration) and space (the present), that is the very mark of its temporal existence. The past can be seen as a dilated present, and the present can be regarded as an extremely contracted form of memory. The actual contracts virtual states within itself, and similarly, the virtual dilates. See Massumi (1992, 63–64).

7 "Against this idea of the absolute originality and unforeseeability of forms our whole intellect rises in revolt. The essential function of our intellect, as the evolution of life has fashioned it, is to be a light for our conduct, to make ready for our action on things, to foresee, for a given situation, the events, favorable or unfavorable, which may follow thereupon. Intellect therefore instinctively selects in a given situation whatever is like something already known: it seeks this out, in order that it may apply its principle that 'like produces like' . . . Like ordinary knowledge in dealing with things science is concerned only with the aspect of *repetition*. Though the whole be original, science will always manage to analyze it into elements or aspects which are approximately a reduction of the past" (CE 29).

8 Bergson's affinity with William James and the tradition of American pragmatism has been well documented. See, for example, Bergson's letter to John Dewey in *The New Bergson* (Mullarkey 1999b, 86). See also James (1970, 1996).

9 Irigaray makes a similar argument about female sexuality and desire when she claims that it is the sex "which is not one," the sex that is not enumerable, countable, and where the distinction between one organ, or orgasm, and another is always artificial or imposed from without. Although Irigaray, to my knowledge, makes no mention of Bergson's writings in her publications, her understanding of the fluid continuity of bodily experience is closely allied with Bergson's understanding of the continuity of the real and the fullness of all forms of life, their qualitative differences from each other. See Irigaray (1985, 1987).

10 Bergson elaborates in "The Perception of Change" in *Creative Mind*: "All real change is an indivisible change. We like to treat it as a series of distinct states which form, as it were, a line in time. That is perfectly natural. If change is continuous in us and also in things, on the other hand, in order that the uninterrupted change, which each of us calls 'me' may act upon the uninterrupted change that we call a 'thing,' these two changes must find themselves, with regard to one another, in a situation like that of the two trains" (CM 172).

11 The Bergsonian philosopher, returning to the fluidity of things, "will see the material world melt back into a simple flux, a continuity of flowing, a becoming. And he will thus prepare to discover real duration there where it is still more useful to find it, in the realm of life and consciousness" (CE 369).

12 "It is true that in the universe itself two opposite movements are to be distinguished . . . 'descent' and 'ascent.' The first only unwinds a roll ready prepared.

In principle it might be accomplished almost instantaneously, like releasing a spring. But the ascending movement, which corresponds to an inner work of ripening or creating, endures essentially, and imposes its rhythms on the first, which is inseparable from it" (CE 11).

13 Luria (1987a) presents a remarkable description and analysis of a man whose cerebral lesions were so severe that he was unable to remember anything but a few of the preceding seconds, a man who lives only in the present as a being entirely at the mercy of utter contingency. In another text, he forcefully describes a unique subject, who seems incapable of forgetting. See Luria (1987b).

14 Bergson's interest in Darwinian thought is there even in his first writings. In *Time and Free Will*, Darwin's work, especially *The Expression of the Emotions* (1965), played a small role. Bergson uses Darwin's discussion of the physiological expression of rage to illustrate his argument that certain affects, especially violent emotions such as "acute desire, uncontrolled anger, passionate love, violent hatred" (TFW 28), which he claims are reducible to muscular and physiological tensions, are linked to processes that are preparatory of action. These emotions, although they may be linked to an idea, a representation, or a reflection, are the outlines of muscular and bodily responses: fighting, fleeing, and so on. The emotions and their motor accompaniments could be construed as an intensification of existing emotions, but, Bergson claims, it would be a mistake to regard acute desire as a more intensive version, a magnification, of a fleeting or minor desire. The intensity of emotions is the effect not of an increase in magnitude but of added and increasingly more complex mixtures of new sensations that accumulate as emotions prolong themselves. These extreme emotions are experienced through the increase in muscular actions and are thus (mis)interpreted as a change in (intensive) magnitude.

15 See, for example, Bergson's objections to the term and the position it implies (CE 42).

16 The essence of mechanical explanation, in fact, is to regard the "future and the past as calculable functions of the present, and thus to claim that *all is given* . . . But radical finalism is quite as unacceptable, and for the same reason. The doctrine of teleology, in its extreme form . . . implies that things and beings merely realize a programme previously arranged. But if there is nothing unforeseen, no invention or creation in the universe, time is useless again. As in the mechanistic hypothesis, here again it is supposed that *all is given.* Finalism thus understood is only inverted mechanism" (CE 37–39).

17 Jacques Monod (1971) is perhaps one of the clearest apologists for a sophisticated mechanism. The gene, or rather genetic structure, contains all the information, stored in advance and developing unmodified, of the organism to come. This organism is a direct realization of the genetic plan, which he specifically likens to architectural plans: "No preformed and complete structure preexisted anywhere; but the architectural plan for it was present in its very constituents. It can

therefore come into being spontaneously and autonomously, without outside help and without the injection of additional information. The necessary information was present, but unexpressed, in its constituents. The epigenetic building of a structure is not a *creation*; it is a *revelation*" (7; emphasis in original).

18 For a more detailed elaboration of the implications of a Bergsonian account for a Dennett-style algorithmic reduction of evolutionary processes to the mindless variation of genetic structure, see Adamson (1999) and Pearson (2002, 79–86).

19 Bergson remains open to the question of the rate of change, a question to which Gould (1989) directs himself in his account of punctuated equilibrium, that is, in his claim that evolutionary change is erratic, remaining stable for long periods, then undergoing rapid transformations in relatively short periods. Bergson's position is agnostic to how rapidly species change, and whether this is a slow, steady rate of change or an erratic, volatile one. His interest is more in how change occurs, the conditions that make it possible, how changing organs and functions are assimilated into the functioning of the individual and the species.

20 "Reality is global and undivided growth, progressive invention, duration: it resembles a gradually expanding rubber balloon assuming at each moment unexpected forms. But our intelligence imagines its origin and evolution as an arrangement and rearrangement of parts which supposedly merely shift from one place to another; in theory therefore, it should be able to foresee any one state of the whole: by positing a definite number of stable elements, one has predetermined all their possible combinations" (*CM* 112–113).

21 "Radical indeed is the difference between an evolution whose continuous phases penetrate one another by a kind of internal growth, and an unfurling whose distinct parts are placed in juxtaposition to one another" (*CM* 20).

22 Bergson's arguments about the evolutionary success of the eye and its privileged position in delaying immediate reactions are elaborated in some detail in Hans Jonas's (1956) analysis of the evolutionary, historical, and phenomenological privileging of vision.

9 Intuition and the Virtual

1 Bergson has no problem in principle with the capacity to simulate or even artificially produce life. Given that some of his writings are now over a century old, his claims, again, seem remarkably prophetic: "That the united efforts of physics and chemistry to manufacture matter resembling living matter may one day be successful is by no means improbable, for life proceeds by insinuating, and the forces which drew matter away from pure mechanism could not have taken hold of matter had it not first adopted mechanism . . . In other words, life must have installed itself in a matter which has already acquired some of the characteristics of life without the work of life" (*ME* 20).

This seems a remarkably prescient description of the work of many contemporary theorists of a-life. See, for example, Langton (1989), Langton, Taylor, Farmer, and Rasmussen (1992), and Levy (1992). For a more detailed discussion of Bergson's possible relevance to theories of a-life, see Grosz (1999).

2 Craig Reynolds's work on the simulation of flocking behavior has intrigued complexity theorists for over fifteen years. Reynolds's concept involved creating computer-animated bird-like entities, 'boids,' in a screen-based spatial environment which was populated by various walls and obstacles. Each boid was programmed using three simple rules:

> 1. It tried to maintain a minimum distance from other objects in the environment, including other boids.
> 2. It tried to match velocities with boids in its neighborhood.
> 3. It tried to move toward the perceived center of the mass of boids in its neighborhood.
>
> What was striking about these rules was that none of them said: "Form a flock." Quite the opposite: the rules were entirely local, referring only to what an individual boid could see and do in its own vicinity. (Waldrop 1992, 241–242)

For more information about these computer programs, see the boids Web site: ⟨http://www.cs.ucl.ac.uk/staff/a.steed/boids.html⟩.

3 For further details on the bottom-up elaboration of complex forms of emergent behavior, in which the units, whether produced by organisms or by computer programming, function without conscious coordination or collusion, see Keller (1996), Gordon (1999), Levy (1992), and Johnson (2001). Deborah Gordon's pioneering research on ant colonies shows that the colony has a life span of around fifteen years, whereas its individual members (except for the queen) live varying lengths of time, but usually no longer than a year; that is, the colony has features and qualities that mark all its members, though it has no specific or universal rules for their interactions. Thus, larger ecological/social units (such as the colony) affect the smaller individual units (each ant), just as its location within an organ structures which elements of the genetic code a particular cell might utilize for its functioning.

4 Bergson goes so far as to suggest that while these more mobile plants are clearly regulated by a tendency that moves from torpor to instinct, there may be a kind of elementary consciousness attributable to moving plants: "Even if we could refer the instincts of animals to habits intelligently acquired and hereditarily transmitted, it is not clear how this sort of explanation could be extended to the vegetable world, where effort is never intelligent, even supposing it is sometimes conscious. And yet, when we see with what sureness and precision climbing plants use their tendrils, what marvelously combined manoeuvres the orchids

perform to procure their fertilization by means of insects, how can we help thinking that these are so many instincts?" (*CE* 170).

5 "For [life] to grow and evolve, there are two ways open. It may take the path towards movement and action,—movement growing ever more effective, action growing freer and freer. The path towards movement involves risk and adventure, but also it involves consciousness, with its growing degrees of intensity and depth. It may take the other path, it may abandon the faculty of acting and choosing, the potentiality of which it carries within it, may accommodate itself to obtain from the spot where it is all it requires for its support, instead of going abroad to seek it . . . These are the two paths which lie open before the evolution of life" (*ME* 12).

6 One cannot ignore the phallic imagery in much of Bergson's analysis of animal existence. It is almost as if the plant/animal distinction represents a distinction between a gentle feminine expenditure and a violent masculine expulsion. These metaphors of explosion, discharge, depletion of energy, and so on that Bergson uses resonate with the imagery of male sexuality. The phallocentrism of Bergson's language is, among other things, the object of analysis of a doctoral dissertation, "Irigaray and Bergson," by Rebecca Hill (Monash University, 2005). See also Olkowski (2000).

7 "Life is tendency, and the essence of a tendency is to develop in the form of a sheaf, creating, by its very growth, divergent directions among which its impetus is divided" (*CE* 99).

8 Bergson proliferates this example with a number of others regarding the apparent insight, or knowledge, of insects with no central nervous system. Here is one detailed example, of the small beetle, Sitaris:

> [The Sitaris] lays its eggs at the entrance of the underground passages dug by a certain kind of bee, the Anthophora. Its larva, after long waiting, springs upon the male Anthophora as it goes out of the passage, clings to it, and remains attached until the "nuptial flight," when it seizes the opportunity to pass from the male to the female, and quietly waits until it lays its eggs. It then leaps on the egg which serves as a support for it in the honey, devours the egg in a few days, and, resting on the shell, undergoes its first metamorphosis. Now organized to float on the honey, it consumes this provision of nourishment, and becomes a nymph, then a perfect insect. Everything happens *as if* the larva of the Sitaris, from the moment it was hatched, knew that the male Anthophora would first emerge from the passage; that the nuptial flight would give it the means of conveying itself to the female, who would take it to a store of honey sufficient to feed it after its transformation . . . And equally all this happens *as if* the Sitaris itself knew that its larva would know all these things. The knowledge, if knowledge there be, is only

> implicit. It is reflected outwardly in exact movements instead of being reflected inwardly in consciousness. It is none the less true that the behavior of the insect involves, or rather, evolves, the idea of definite things existing or being produced in definite points of space and time, which the insect knows without having learned them. (CE 146–147; see CE 172–173 for other examples)

9 It takes us longer to change ourselves than to change our tools. Our individual and even social habits survive a good while the circumstances for which they were made, so that the ultimate effects of an invention are not observed until its novelty is already out of sight. A century has elapsed since the invention of the steam-engine, and we are only just beginning to feel the depths of the shock it gave us. But the revolution it has effected in industry has nevertheless upset human relations altogether. New ideas are arising, new feelings are on the way to flower. In thousands of years . . . our wars and our revolutions will count for little, even supposing they are remembered at all; but the steam-engine, and the procession of inventions of every kind that accompanied it, will perhaps be spoken of as we speak of the bronze or of the chipped stone of pre-historic times: it will serve to define an age. (CE 138–139)

10 For Bergson, the tool not only transforms the material world in a potentially endless fashion, it also transforms its user: "For every need that it satisfies, it creates a new need; and so, instead of closing, like the instinct, the round of action within which the animal tends to move automatically, it lays open to activity an unlimited field into which it is driven further and further, and made more and more free. But this advantage of intelligence over instinct only appears at a late stage, when intelligence, having raised construction to a higher degree, proceeds to construct constructive machinery" (CE 141).

11 For example, see Gunter, "Bergson and the War against Nature," in Mullarkey (1999a).

12 "Thus the instinctive knowledge which one species possesses of another on a certain particular point has its root in the very unity of life, which is, to use the expression of an ancient philosopher, a 'whole sympathetic to itself.' It is impossible to consider some of the special instincts of the animal and of the plant, evidently arisen in extraordinary circumstances, without relating them to those recollections, seemingly forgotten, which spring up suddenly under the pressure of an urgent need" (CE 167).

13 "The new-born child, so far as being intelligent, knows neither definite objects nor a definite property of any object; but when, a little later on, he will hear an epithet being applied to a substantive, he will immediately understand what it means. The relation of attribute to subject is therefore seized by him naturally, and the same might be said of the general relation expressed by the verb, a

relation so immediately conceived by the mind that language can leave it to be understood . . . Intelligence, therefore, naturally makes use of relations of like with like, of content to container, of cause to effect, etc. which are implied in every phrase in which there is a subject, an attribute and a verb" (CE 148).

14 If language is the mode that interests contemporary consciousness in its self-reflections, it is geometry, also an innate form of consciousness, also considered an innate knowledge of spatial relations, that preoccupied the ancients.

15 "Whatever be the force that is at work in the genesis of the nervous system of the caterpillar, to our eyes and intelligence it is only a juxtaposition of nerves and nervous centres. It is true that we thus get the whole outer effect of it. The Ammophila [a wasp], no doubt, discerns but very little of that force, just what concerns itself; but at least it discerns it from within, quite otherwise than by a process of knowledge—by an intuition (*lived* rather than *represented*), which is probably like what we call divining sympathy" (CE 175).

16 Bergson presents a favorite example: sketches or photographs of the city of Paris are analytic fragments, which, no matter how many one makes, or how closely put together they are side by side, will never give one a sense of the city of Paris if one has never been there. But if one has been there, one has experienced these relations represented in the sketches in an immediacy. This makes it easy to reconstruct and place various sketches side by side in a way that would be impossible without this experience (incidentally, this may explain why viewing someone else's photographs is an unfailingly boring experience, while viewing one's own is inherently interesting!): "Even with an infinity of sketches as exact as you like, even with the word 'Paris' to indicate that they must bear close connection, it is impossible to travel back to an intuition one has not had, and gain the impression of Paris if one has never seen Paris" (CM 201). No matter how many snapshots of Paris one adds together, the simplicity, unity, and cohesion of one's experience of the city is only evoked but never captured. Intuition begins, not with the collection of snapshots, immobile elements that, when added together, are supposed to simulate the unity of the event itself, but with the mobile whole, with its movement of dynamic change. Anthony Wilden's (1972) distinction between digital and analogue seems to capture Bergson's argument here.

17 Intuition is a remarkably difficult activity, for it casts us into the unknown, where we cannot utilize either habits or reason to anticipate what is to come, and where we will feel the most vulnerable. This is why intuition is intermittent, even if what it intuits is continuous. It can occur only from moment to moment: "To reason on abstract ideas is easy; metaphysical construction is only a game, however slightly one is predisposed to it. To penetrate the mind intuitively is perhaps more painful, but no philosopher will work at it for long at a stretch. He will quickly perceive each time what he is capable of perceiving" (CM 41).

18 "Analysis operates on immobility, while intuition is located in mobility or, what amounts to the same thing, in duration. That is the very clear line of demarca-

tion between intuition and analysis. One recognizes the real, the actual, the concrete, by the fact that it is variability itself. One recognizes the element by the fact that it is invariable. And it is invariable by definition, being a schema, a simplified reconstruction, often a mere symbol, in any case, a view taken of the reality that flows" (CM 213).

19 For example, in his research on computer viruses, Eugene Spafford (1992) argues that, as far as the properties a-life scientists want to associate with the definition of life, a computer virus could qualify as life. Yet, he concludes that this indicates a problem with the definition of life on which computer programs have relied. He cites the following criteria, which he believes the computer virus satisfies:

1. Life is a pattern in space-time rather than a specific material object.
2. Self-reproduction, in itself or in a related organism.
3. Information storage of a self-representation.
4. A metabolism that converts matter/energy.
5. Functional interactions with the environment.
6. Interdependence of parts.
7. Stability under perturbations of the environment.
8. The ability to evolve.
9. Growth or expansion. (741)

Noting, indeed arguing in favor of the analogy, Spafford is clearly disturbed at the implication that one may not be able to draw a hard and fast line dividing life from its simulation. At the very end of his paper he reasserts in somewhat dogmatic form exactly that which his own paper has helped problematize. He argues that "we must never lose sight of the fact that 'real life' is of much more importance than 'artificial life,' and we should not allow our experiments to threaten our experimenters" (744), when in fact his own arguments make it increasingly difficult to divide the experimenter from the experiment. For more on the links between biological and computer viruses, see Moya, Domingo, and Holland (1995), Dawkins (1986), and Levy (1992).

20 Langton (1989, 2) makes explicit the relative indifference of biological concepts of life to the specific modes of materiality to which it has up to now been confined: "Certainly life, as a dynamic physical process, could haunt other physical material: the material just needs to be organized in the right way. Just as certainly, the dynamic processes that constitute life—in whatever material bases they might occur—must share certain universal features—features that allow us to recognize life by its dynamic *form* alone, without reference to its *matter.* This *general* phenomenon of life—life writ-large across all possible material substrates—is the true subject matter of biology."

Whenever the specificities of matter are regarded merely as "substrate," "ground," or "support" of a program, form, or idea, it seems that we return to the reign of Platonism and the profound somatophobia to which it gave rise. It is significant that Langton's conception of matter as substrate has been challenged, not entirely surprisingly, from the point of view of molecular biology, "according to which 'form' and 'matter' do not represent separate realms" (Emmeche 1992, 466).

Conclusion

1 For a more contemporary example, see Hawking (1988, 1993).
2 For further details on the question of complexity, see Coveney and Highfield (1995, 9), Johnson (2001), Waldrop (1992), Barbour (2000), and Langton (1989).
3 Stable and unstable structures can be distinguished, basically, in terms of their modes of transformation, their temporal movement. Stable dynamic systems are those in which a slight change in initial conditions produces correspondingly slight effects. Newtonian systems, while approaching the status of universal laws, exhibit this general stability: our real-world measurements are at best approximations, but their approximative character seems enough to assure us of their general validity in so-called midsize objects. Unstable dynamic systems, systems on the edge of chaos, are those in which small fluctuations in initial conditions are amplified over the course of time to produce major effects. These systems are known as chaotic, dissipative, or nonequilibrium systems. These may be deterministic systems whose particular outcome is nevertheless unpredictable as a result of the complexity of calculations predictability entails. See Prigogine (1997, 30).
4 In some sense, such research has been anticipated in Foucault's genealogical analyses of present institutions (the prison, the hospital, the school) as well as in discourses (true knowledges, the "human sciences," psychiatry, criminology, sexology) that seek the lines of fortification of power's operation in order to discover the sites of resistance inevitably produced, sites that reveal present points of fragility. See Foucault (1977, 1978, 1979a, 1979b).

REFERENCES

Adamson, Gregory Dale. 1999. "Henri Bergson: Time, Evolution and Philosophy." *World Futures* 54.1: 135–162.

——. 2000. "Science and Philosophy: Two Sides of the Absolute." *Pli: The Warwick Journal of Philosophy* 9: 53–86.

——. 2002. *Philosophy in the Age of Science and Capital.* London: Continuum.

Alexander, Ian. 1957. *Bergson: Philosopher of Reflection.* London: Bowes and Bowes.

Antliff, Mark. 1993. *Inventing Bergson: Cultural Politics and the Parisian Avant-Garde.* Princeton, N.J.: Princeton University Press.

Bachelard, Gaston. 2000. *The Dialectic of Duration.* Trans. M. MacAllister Jones. Manchester, England: Clinamen Press.

Barbour, Julian P. 2000. *The End of Time: The Next Revolution in Physics.* Oxford: Oxford University Press.

Barrish, Phillip. 1991. "Accumulating Variation: Darwin's *On the Origin of Species* and Contemporary Literary and Cultural Theory." *Victorian Studies* (summer): 431–453.

Bataille, Georges. 1985. *Visions of Excess: Selected Writings 1927–1939.* Trans. Allen Stoekl. Manchester, England: Manchester University Press.

Benton, Ted. 1995. "Science, Ideology and Culture: Malthus and *The Origin of Species.*" In *Charles Darwin's The Origin of Species: New Interdisciplinary Essays*, ed. David Amigone and Jeff Wallace, 68–94. Manchester, England: Manchester University Press.

Bergson, Henri. 1921. *Mind-Energy.* Trans. H. Wilden Carr. London: Macmillan.

——. 1944. *Creative Evolution.* Trans. Arthur Mitchell. New York: Modern Library.

——. 1958. *The World of Dreams.* Trans. Wade Baskin. New York: Philosophical Library.

——. 1959. *Time and Free Will: An Essay on the Immediate Data of Consciousness.* Trans. F. L. Pogson. London: George Allen and Unwin.

——. 1965. *Duration and Simultaneity with Reference to Einstein's Theory.* Trans. Leon Jacobson. Indianapolis: Bobbs-Merrill. (Orig. pub. 1922.)

——. 1986. *The Two Sources of Morality and Religion.* Trans R. Ashley Audra and Cloudsley Brereton. Notre Dame, Ind.: University of Notre Dame Press. (Orig. pub. 1932.)

——. 1988. *Matter and Memory.* Trans N. M. Paul and W. S. Palmer. New York: Zone. (Orig. pub. 1896.)

——. 1992. *The Creative Mind: An Introduction to Metaphysics.* Trans. Mabelle L. Andison. New York: Citadel Press. (Orig. pub. 1919.)

Berkowitz, Peter. 1995. *Nietzsche: The Ethics of an Immoralist.* Cambridge, Mass.: Harvard University Press.

Blackwell, Antoinette Brown. 1875. *The Sexes throughout Nature.* New York: G.P. Putnam's Sons.

Blondel, Eric. 1985. "Nietzsche: Life as Metaphor." In *The New Nietzsche: Contemporary Styles of Interpretation*, ed. D. B. Allison, 150–175. Cambridge, Mass.: MIT Press.

——. 1991. *Nietzsche: The Body and Culture. Philosophy as a Philological Genealogy.* Trans. Seàn Hand. Stanford: Stanford University Press.

Boden, Margaret A., ed. 1996. *The Philosophy of Artificial Life*. Oxford: Oxford University Press.

Bogue, Ronald. 2003. *Deleuze on Cinema.* New York: Routledge.

Boundas, Constantin V., ed. 1993. *The Deleuze Reader.* New York: Columbia University Press.

——. 1996. "Bergson-Deleuze: An Ontology of the Virtual." In *Deleuze: A Critical Reader*, ed. P. Patton. Oxford: Blackwell.

Braidotti, Rosi. 1994. *Nomadic Subjects: Embodiment and Sexual Difference in Contemporary Feminist Theory.* New York: Columbia University Press.

——. 2002. *Metamorphoses: Towards a Materialist Theory of Becoming.* Oxford: Polity Press.

Buchanan, Ian, ed. 1997. *A Deleuzian Century*. Special issue of *Southern Atlantic Quarterly* 96.

Butler, Samuel. 1887. *Luck, or Cunning, as the Main Means of Organic Modification? An Attempt to Throw Additional Light upon Darwin's Theory of Natural Selection.* London: Trübner.

Camerini, Jane, ed. 2002. *The Alfred Russel Wallace Reader.* Baltimore: Johns Hopkins University Press.

Čapek, Milič. 1970. *Bergson and Modern Physics.* Dordrecht: Martinus Nijhoff.

——, ed. 1976. *The Concepts of Space and Time: Their Structure and Development.* Boston Studies in the Philosophy of Science. Dordrecht: D. Reidel.

——. 1991. *The New Aspects of Time: Its Continuities and Novelties. Selected Papers in the Philosophy of Science.* Dordrecht: Kluwer Academic.

Cariou, Marie. 1999. "Bergson and the Keyboards of Forgetting." In *The New Bergson*, ed. John Mullarkey, 99–117. Manchester, England: University of Manchester Press.

Casti, John L. 1990. *Searching for Certainty: What Scientists Can Know about the Future.* New York: William Morrow.

——. 1994. *Complexification: Explaining a Paradoxical World through the Science of Surprise.* New York: HarperCollins.

Cavalli-Sforza, Luigi Luca. 2000. *Genes, People and Languages.* Trans. Mark Seielstad. Berkeley: University of California Press.

Cheah, Pheng. 1996. "Mattering." *Diacritics* 26.1: 108–139.

——. 2003. *Spectral Nationality: Passages of Freedom from Kant to Postcolonial Literatures of Liberation.* New York: Columbia University Press.

Christensen, F. M. 1993. *Space-like Time: Consequences of, Alternatives to, and Arguments Regarding the Theory That Time Is Like Space.* Toronto: University of Toronto Press.

Cohen, Jack, and Ian Stewart. 1994. *The Collapse of Chaos: Discovering Simplicity in a Complex World.* Harmondsworth, England: Penguin.

Coveney, Peter, and Roger Highfield. 1995. *Frontiers of Complexity: The Search for Order in a Chaotic World.* New York: Fawcett Columbine.

Crary, Jonathan, and Sanford Kwinter, eds. 1992. *Incorporations.* New York: Zone.

Danto, Arthur C. 1968. *Analytical Philosophy of Knowledge.* New York: Macmillan.

Darwin, Charles. 1885. *The Variation of Animals and Plants under Domestication.* 2 vols. London: J. Murray.

——. 1965. *The Expression of Emotions in Man and Animals.* Chicago: University of Chicago Press. (Orig. pub. 1872.)

——. 1981. *The Descent of Man, and Selection in Relation to Sex.* 2 vols. in one, each vol. separately paginated. Princeton, N.J.: Princeton University Press. (Repr. of 1871 edition, London: J. Murray.)

——. 1996. *The Origin of Species.* Oxford: Oxford University Press. (Orig. pub. 1859.)

Dawkins, Richard. 1986. *The Blind Watchmaker.* New York: Norton.

——. 1989. *The Selfish Gene.* Oxford: Oxford University Press.

de Landa, Manuel. 1991. *War in the Age of Intelligent Machines.* New York: Zone.

——. 1993. "Inorganic Life and Predatory Machines." In *Culture Lab*, ed. B. Boigon, 27–38. New York: Princeton Architectural Press.

——. 1997. *A Thousand Years of Non-Linear History.* New York: Zone.

——. 1999. "Deleuze, Diagrams and the Open-ended Becoming of the World." In *Becomings: Explorations in Time, Memory and Futures*, ed. Elizabeth Grosz, 29–41. Ithaca, N.Y.: Cornell University Press.

——. 2002. *Intensive Science and Virtual Philosophy.* London: Continuum.

Deleuze, Gilles. 1983. *Nietzsche and Philosophy.* Trans. Hugh Tomlinson. London: Athlone Press.

——. 1988. *Bergsonism.* Trans. Hugh Tomlinson and Barbara Habberjam. New York: Zone.

——. 1989. *Cinema 2: The Time-Image.* Trans. Hugh Tomlinson and Robert Galeta. Minneapolis: University of Minnesota Press.

——. 1991. *Empiricism and Subjectivity: An Essay on Hume's Theory of Human Nature.* Trans. Constantin Boundas. New York: Columbia University Press.

——. 1994. *Difference and Repetition.* Trans. Paul Patton. New York: Columbia University Press.

——. 1997. *Essays Critical and Clinical.* Trans. Daniel W. Smith and Michael A. Greco. Minneapolis: Unversity of Minnesota Press.

——. 1999. "Bergson's Conception of Difference." In *The New Bergson*, ed. John Mullarkey, 42–66. Manchester, England: Manchester University Press.

——. 2001. *Pure Immanence: Essays on a Life.* Trans. Anne Boyman. New York: Zone.

Deleuze, Gilles, and Félix Guattari. 1987. *A Thousand Plateaus: Capitalism and Schizophrenia, Vol 2.* Trans. Brian Massumi. Minneapolis: University of Minnesota Press.

Deleuze, Gilles, and Claire Parnet. 1987. *Dialogues.* Trans. Hugh Tomlinson and Barbara Habberjam. London: Athlone Press.

Dennett, Daniel. 1996. *Darwin's Dangerous Idea: Evolution and the Meaning of Life.* New York: Touchstone.

Depew, David, and Bruce H. Weber. 1997. *Darwinism Evolving: Systems Dynamics and the Genealogy of Natural Selection*. Cambridge: MIT Press.

Derrida, Jacques. 1982. "*Ousia and Grammé:* Note on a Note on *Being and Time.*" In *Margins of Philosophy*, trans. Alan Bass. Chicago: University of Chicago Press.

——. 1992. *Given Time: I. Counterfeit Money.* Chicago: University of Chicago Press.

——. 1995a. *The Gift of Death.* Trans. David Wills. Chicago: University of Chicago Press.

——. 1995b. "The Time Is Out of Joint." In *Deconstruction is/in America: A New Sense of the Political*, ed. A. Haverkamp, 14–40. New York: New York University Press.

Dobzhansky, Theodosius. 1982. *Genetics and the Origin of Species.* New York: Columbia University Press. (Orig. pub. 1937.)

Douglass, Paul. 1992. "Deleuze and the Endurance of Bergson." *Thought* 67.264: 47–61.

——. 1998. "Deleuze, Cinema, Bergson." *Social Semiotics* 8.1: 25–35.

Drogould, Alexis, Jacques Ferber, Bruno Corbara, and Dominique Fresneau. 1992. "A Behavioral Simulation Model for the Study of Emergent Social Structures." In *Toward a Practice of Autonomous Systems: Proceedings of the First European Conference on Artificial Life*, ed. Francisco J. Varela and Paul Bourgine, 123–132. Cambridge Mass.: MIT Press.

Dyson, George B. 1997. *Darwin among the Machines: The Evolution of Global Intelligence.* Reading, Mass.: Perseus.

Emmeche, Claus. 1992. "A Semiotical Reflection on Biology, Living Signs and Artificial Life." *Biology and Philosophy* 6: 325–340.

——. 1994. *The Garden in the Machine: The Emerging Science of Artificial Life.* Princeton, N.J.: Princeton University Press.

Erskine, Fiona. 1995. "The Origin of Species and the Science of Female Inferiority." In *Charles Darwin's The Origin of Species: New Interdisciplinary Essays,* ed. David Amigone and Jeff Wallace. Manchester, England: Manchester University Press.

Fausto-Sterling, Anne, Patricia Adair Gowaty, and Marlene Zuk. 1997. "Evolutionary Psychology and Darwinian Feminism." *Feminist Studies* 23.2 (summer): 318–403.

Fletcher, R. P., C. Cannings, and P. G. Blackwell. 1995. "Modelling Foraging Behaviour of Ant Colonies." In *Advances in Artificial Life: Third European Conference on Artificial Life*, ed. F. Morán, A. Moreno, J. J. Merelo, and P. Chacón. Berlin: Springer-Verlag.

Foucault, Michel. 1970. *The Order of Things: An Archaeology of the Human Sciences.* Trans. Alan Sheridan. London: Tavistock.

——. 1972. *The Archaeology of Knowledge.* Trans. Alan M. Sheridan Smith. New York: Harper Colophon.

——. 1977. *Discipline and Punish: The Birth of the Prison.* Trans. Alan Sheridan. London: Allen Lane.

——. 1978. *The History of Sexuality. Volume 1: An Introduction.* Trans. Robert Hurley. London: Allen Lane.

——. 1979a. "Governmentality." Trans. Colin Gordon. *Ideology and Consciousness* 6 (autumn): 5–21.

——. 1979b. "The Life of Infamous Men." In *Michel Foucault: Power, Truth, Strategy.* Ed. M. Morris and P. Patton. Sydney: Feral Publications.

Freud, Sigmund. 1905. *The Interpretation of Dreams.* In *The Standard Edition of the Complete Psychological Works of Sigmund Freud*, ed. James Strachey, vol. 5. Oxford: Hogarth Press.

Game, Ann. 1995. "Time, Space, Memory with Reference to Bachelard." In *Global Modernities*, ed. Mike Featherstone, Scott Lash, and Roland Robertson, 192–208. London: Sage.

Gell-Mann, Murray. 1994. "Complex Adaptive Systems." In *Complexity: Metaphors, Models and Reality*, ed. George A. Cowan, David Pines, and David Meltzer, 17–45. Reading, Mass.: Addison-Wesley.

Gilbert, Nigel, and Rosaria Conte, eds. 1995. *Artificial Societies: The Computer Simulation of Social Life.* London: UCL Press.

Godfrey-Smith, Peter. 1994. "Spencer and Dewey on Life and Mind." In *Artificial Life IV: Proceedings of the Fourth International Workshop on the Synthesis and Simulation of Living Systems*, ed. Rodney A. Brooks and Pattie Maes, 80–89. Cambridge, Mass.: MIT Press.

Gordon, Deborah. 1999. *Ants at Work: How an Insect Society Is Organized.* New York: Free Press.

Gould, Stephen Jay. 1989. *Wonderful Life: The Burgess Shale and the Nature of History.* New York: Norton.

——. 1991. *Bully for Brontosaurus.* New York: Norton.

——. 2002. *The Structure of Evolutionary Theory.* Cambridge, Mass.: Belknap Press of Harvard University Press.

Gowaty, Patricia Adair. 1995. "Battles of the Sexes and Origins of Monogamy." In *Partnerships in Birds*, ed. J. L. Black, 21–52. Oxford: Oxford University Press.

——. 1997a. "Darwinian Feminists and Feminist Evolutionists." In *Feminism and Evolutionary Biology: Boundaries, Intersections, and Frontiers*, ed. P. A. Gowaty, 1–17. New York: Chapman and Hall.

——, ed. 1997b. *Feminism and Evolutionary Biology: Boundaries, Intersections, and Frontiers.* New York: Chapman and Hall.

Griffin, Donald R. 1992. *Animal Minds.* Chicago: University of Chicago Press.

Grosz, Elizabeth. 1994. *Volatile Bodies: Toward a Corporeal Feminism.* Bloomington: Indiana University Press.

——. 1995. *Space, Time and Perversion: Essays on the Politics of Bodies.* New York: Routledge.

——. 1996. "Intolerable Ambiguity: Freaks as/at the Limit." In *Freakmaking: Constituting Corporeal and Cultural Others*, ed. R. G. Thomson, 55–65. New York: New York University Press.

——, ed. 1999. *Becomings: Essays in Time, Memory, and Futures.* Ithaca, N.Y.: Cornell University Press.

——. 2001. *Architecture from the Outside: Essays on Virtual and Real Space.* Cambridge, Mass.: MIT Press.

Grünbaum, Adolf. 1973. *Philosophical Problems of Space and Time.* Dordrecht: D. Reidel.

Gunn, Alexander. 1920. *Bergson and His Philosophy.* London: Methuen.

Gunter, Pete A. Y. 1969. *Bergson and the Evolution of Physics.* Knoxville: University of Tennessee Press.

——. 1971. "Bergson's Theory of Matter and Modern Cosmology." *Journal of the History of Ideas* 32.4: 525–543.

Guth, Alan H., and Alan P. Lightman. 1998. *The Inflationary Universe: Quest for a New Theory of Cosmic Origin.* Reading, Mass.: Perseus Press.

Hamilton, William. 1964. "The Genetic Evolution of Social Behavior." *Journal of Theoretical Biology* 7 (part 1): 1–16; (part 2): 17–52.

Hawking, Stephen. W. 1988. *A Brief History of Time: From the Big Bang to Black Holes.* New York: Bantam.

——. 1993. *Black Holes and Baby Universes and Other Essays.* London: Bantam.

Heidegger, Martin. 1979. *Nietzsche.* Vol. 1. Trans. David Farrell Krell. San Francisco: Harper and Row.

——. 1984. *Nietzsche.* Vol. 2. Trans David Farrell Krell. San Francisco: Harper and Row.

Heims, Steve. 1980. *John Von Neumann and Norbert Wiener: From Mathematics to the Technologies of Life and Death.* Cambridge, Mass.: MIT Press.

Heller, Erich. 1988. *The Importance of Nietzsche: Ten Essays.* Chicago: University of Chicago Press.

Herken, Rolf, ed. 1994. *The Universal Turing Machine: A Half-Century Survey.* Vienna: Springer-Verlag.

Hill, Rebecca. 2004. "Irigaray and Bergson." Ph.D. diss., Critical Theory Program, Monash University.

Hodges, Andrew. 1983. *Alan Turing, the Enigma.* London: Vintage.

——. 1994. "Alan Turing and the Turing Machine." In *The Universal Turing Machine: A Half-Century Survey*, ed. R. Herken, 3–15. Vienna: Springer-Verlag.

Huxley, Julian T. 1926. *Essays of a Biologist.* London: Chatto and Windus.

——. 1942. *Evolution: The Modern Synthesis.* London: Allen and Lane.

Irigaray, Luce. 1985. *This Sex Which Is Not One.* Trans. Catherine Porter with Carolyne Burke. Ithaca, N.Y.: Cornell University Press.

——. 1987. "Is the Subject of Science Sexed?" *Hypatia* 2: 65–87.

——. 1991. *Marine Lover: Of Friedrich Nietzsche.* Trans. Gillian C. Gill. New York: Columbia University Press.

——. 1996. *I Love to You: Sketch of a Possible Felicity in History.* New York: Routledge.

Jakobson, Roman, and Morris Halle. 1956. *The Fundamentals of Language.* The Hague: Mouton.

James, William. 1970. *Pragmatism and Four Essays from The Meaning of Truth.* Cleveland: Meridian.

——. 1996. *A Pluralistic Universe: Hibbert Lectures at Manchester College on the Present Situation of Philosophy.* Lincoln: University of Nebraska Press.

Johnson, Steven. 2001. *Emergence: The Connected Lives of Ants, Brains, Cities and Software.* New York: Scribner's.

Jonas, Hans. 1956. "The Nobility of Sight." In *The Phenomenon of Life: Toward a Philosophical Biology.* New York: Harper and Row.

Judson, Horace Freeland. 1979. *The Eighth Day of Creation: Makers of the Revolution in Biology.* London: Penguin.

Kampis, George, ed. 1992. *Self-Modifying Systems: A New Framework for Dynamics, Information and Complexity.* London: Pergamon.

——. 1993. "Creative Evolution." *World Futures* no. 38: 131–137.

——. 1995. "The Inside and Outside Views of Life." In *Advances in Artificial Life: Third European Conference on Artificial Life*, ed. F. Morán, A. Moreno, J. J. Merelo, and P. Chacón. Berlin: Springer-Verlag.

Kampis, George, and P. Weibel, eds. 1992. *Endophysics: A New Approach to the Observer-Problem with Applications in Physics, Biology and Mathematics.* Santa Cruz: Aerial Press.

Kaufman, Stuart. 1993. *The Origins of Order: Self-Organization and Selection in Evolution.* New York: Oxford University Press.

Keller, Evelyn Fox. 1996. "The Force of the Pacemaker Concept in Theories of Aggregation in Cellular Slime Mold." In *Reflections on Gender and Science.* New Haven: Yale University Press.

——. 1998. "Structures of Heredity." *Biology and Philosophy* no. 13.

Klossowski, Pierre. 1985. "Nietzsche's Experience of the Eternal Return." In *The New Nietzsche: Contemporary Styles of Interpretation*, ed. David B. Allison, 107–120. Cambridge, Mass.: MIT Press.

——. 1997. *Nietzsche and the Vicious Circle.* Trans. Daniel W. Smith. Chicago: University of Chicago Press.

Kristeva, Julia. 1984. *The Revolution in Poetic Language.* New York: Columbia University Press.

Kuhn, Thomas. 1970. *The Structure of Scientific Revolutions.* Chicago: University of Chicago Press.

Kwinter, Sanford. 1992. "Emergence: Or the Artificial Life of Space." In *Anywhere*, ed. Cynthia Davidson, 164–171. New York: Rizzoli.

——. 1993. "Soft Systems." In *Culture Lab*, ed. B. Boigon, 207–228. New York: Princeton Architectural Press.

——. 2001. *Architectures of Time: Toward a Theory of the Event in Modernist Culture.* Cambridge, Mass.: MIT Press.

Lacey, A. R. 1986. *Bergson.* London: Routledge.

Lamarck, Jean-Baptiste. 1914. *The Zoological Philosophy.* London: Macmillan.

Lampert, Lawrence. 1986. *Nietzsche's Teaching.* New Haven: Yale University Press.

Langton, Christopher, ed. 1989. *Artificial Life*. Redwood City, Calif.: Addison-Wesley.

Langton, Christopher, C. Taylor, J. D. Farmer, and S. Ramussen, eds. 1992. *Artificial Life II.* Redwood City, Calif.: Addison-Wesley.

Leroi-Gourhan, André. 1993. *Gesture and Speech.* Trans. Anna Bostock Berger. Cambridge, Mass.: MIT Press.

Lestienne, Rémy. 1998. *The Creative Power of Chance.* Urbana: University of Illinois Press.

Levinas, Emmanuel. 1987. *Time and the Other.* Pittsburgh: Dusquesne University Press.

Levy, Steven. 1992. *Artificial Life: The Quest for a New Creation.* New York: Pantheon.

Lewontin, Richard C. 1992. *Biology as Ideology: The Doctrine of* DNA. New York: HarperCollins.

Lewontin, Richard C., S. Rose, and L. J. Kamin. 1984. *Not in Our Genes: Biology, Ideology and Human Nature.* New York: Pantheon.

Lieb, Irwin C. 1991. *Past, Present and Future: A Philosophical Essay about Time.* Urbana: University of Illinois Press.

Linde, Andrei. 1994. "The Self-Reproducing Inflationary Universe." *Scientific American* (November): 48–55.

Lingis, Alphonso. 1985. "The Will to Power." In *The New Nietzsche*, ed. David B. Allison, 37–63. Cambridge, Mass.: MIT Press.

Luria, A. R. . 1987a. *The Man with a Shattered World: The History of a Brain Wound.* Trans. Lynn Solotaroff. Cambridge, Mass.: Harvard University Press.

——. 1987b. *The Mind of a Mnemonist: A Little Book about a Vast Memory.* Trans. Lynn Solotaroff. Cambridge, Mass.: Harvard University Press.

Lyell, Charles. 1863. *The Geological Evidences of the Antiquity of Man, with Remarks on the Theories of the Origin of Species by Variation.* London: John Murray.

Malthus, Thomas. 1976. *An Essay on the Principle of Population, as It Affects the Future Improvement of Society, with Remarks on the Speculations of Mr. Goodwin, M. Condorcet and Other Writers.* London: Norton. (Orig, pub. 1798.)

Marx, Karl. 1974. *Capital: A Critical Analysis of Capitalist Production.* Vols. 1–3. Trans. Samuel Moore and Edward Aveling. Moscow: Progress.

Massumi, Brian. 1992. *A Reader's Guide to Capitalism and Schizophrenia.* Minneapolis: University of Minnesota Press.

——. 2002. *Parables for the Virtual: Movement, Affect, Sensation.* Durham, N.C.: Duke University Press.

Maturana, Victor, R. Humberto, and Francisco J. Varela. 1980. *Autopoesis and Cognition: The Realization of the Living* . Boston Studies in the Philosophy of Science, vol. 42. Dordrecht: D. Reidel.

Maxwell, Donald R. 1999. *The Abacus and the Rainbow: Bergson, Proust and the Digital-Analogic Opposition.* New York: Peter Lang.

Mayr, Ernst. 1997. *Evolution and the Diversity of Life: Selected Essays.* Cambridge, Mass.: Belknap Press of Harvard University Press.

Merleau-Ponty, Maurice. 1962. *The Phenomenology of Perception.* Trans. Colin Smith. London: Routledge and Kegan-Paul.

——. 1963. *The Primacy of Perception.* Trans. James Edie. Evanston, Ill.: Northwestern University Press.

——. 1964. *Signs.* Trans. Richard C. McCleary. Evanston, Ill.: Northwestern University Press.

——. 1968. *The Visible and the Invisible.* Trans. Alphonso Lingis. Evanston, Ill.: Northwestern University Press.

——. 1970. *In Praise of Philosophy and Other Essays.* Trans. John Wild, James Edie, and John O'Neill. Evanston, Ill.: Northwestern University Press.

——. 1983. *The Structure of Behavior.* Trans. Alden L. Fisher. Pittsburgh: Duquesne University Press.

Miller, S., and L. E Orgel. 1974. *The Origins of Life on Earth.* Englewood Cliffs, N.J.: Prentice-Hall.

Monod, Jacques. 1971. *Chance and Necessity: An Essay on the Natural Philosophy of Modern Biology.* Harmondsworth, England: Penguin.

Morgan, Elaine. 1992. "Darwin and Feminism." *College Anthropology* 15.1: 45–52.

Moya, Andrés, Esteban Domingo, and John J. Holland. 1995. "RNA Viruses: A Bridge between Life and Artificial Life." In *Advances in Artificial Life: Third European*

Conference on Artificial Life, ed. F. Morán, A. Moreno, J. J. Merelo, and P. Chacón. Berlin: Springer-Verlag.

Mullarkey, John C. 1994. "Duplicity in the Flesh: Bergson and Current Philosophy of the Body." *Philosophy Today* (winter): 339–355.

——. 1999a. *Bergson and Philosophy.* Notre Dame, Ind.: Notre Dame University Press.

——, ed. 1999b. *The New Bergson.* Manchester, England: Manchester University Press.

Nehemas, Alexander. 1985. *Nietzsche: Life as Literature.* Cambridge, Mass.: Harvard University Press.

Newton, Isaac. 1999. *The Principia Mathematica: Principles of Natural Philosophy.* Trans. I. Bernard Cohen and Anne Whitman. Berkeley: University of California Press. (orig. pub. 1687.)

Nietzsche, Friedrich. 1965. *Thus Spoke Zarathustra: A Book for All and None.* Trans. Walter Kaufmann. Harmondsworth, England: Penguin.

——. 1967. *On the Genealogy of Morals and Ecce Homo.* Trans. Walter Kaufmann. New York: Vintage.

——. 1968a. *Twilight of the Idols and The Anti-Christ.* Trans. R. J Hollingdale. Harmondsworth, England: Penguin.

——. 1968b. *The Will to Power.* Trans. Walter Kaufmann and R. J. Hollingdale. New York: Vintage.

——. 1972. *Beyond Good and Evil: Prelude to a Philosophy of the Future.* Trans. R. J. Hollingdale. Harmondsworth, England: Penguin.

——. 1974. *The Gay Science.* Trans. Walter Kaufmann. New York: Vintage.

——. 1995. "On the Utility and Liability of History for Life." In *Unfashionable Observations.* Trans. Richard T. Gray. Stanford: Stanford University Press.

Olkowski, Dorothea. 2000. "The End of Phenomenology: Bergson's Interval in Irigaray." *Hypatia* 15.3.

Oyama, Susan. 2000a. *Evolution's Eye: A Systems View of the Biology-Culture Divide.* Durham, N.C.: Duke University Press.

——. 2000b. *The Ontogeny of Information: Developmental Systems and Evolution.* Durham, N.C.: Duke University Press.

Pearson, Keith Ansell, ed. 1997a. *Deleuze and Philosophy: The Difference Engineer.* London: Routledge.

——. 1997b. *Viroid Life: Perspectives on Nietzsche and the Transhuman Condition.* London: Routledge.

——. 1999. *Germinal Life: The Difference and Repetition of Deleuze.* London: Routledge.

——. 2002. *Philosophy and the Adventure of the Virtual: Bergson and the Time of Life.* London: Routledge.

Penrose, Roger. 1995. *Shadows of the Mind.* London: Vintage.

Prigogine, Ilya. 1997. *The End of Certainty: Time, Chaos and the New Laws of Nature.* New York: Free Press.

Prigogine, Ilya, and Isabelle Stengers. 1984. *Order Out of Chaos: Man's New Dialogue with Nature.* London: HarperCollins.

Rajchman, John. 1992. "Anywhere and Nowhere." In *Anywhere*, ed. C. Davidson, 228–233. New York: Rizzoli.

——. 1998. *Constructions.* Cambridge, Mass.: MIT Press.

——. 2000. *The Deleuze Connections.* Cambridge, Mass.: MIT Press.

Ricardo, David. 1817. *Principles of Political Economy and Taxation.* London: John Murray.

Rodowick, David N. 1997. *Gilles Deleuze's Time Machine.* Durham, N.C.: Duke University Press.

Rose, Michael R. 1998. *Darwin's Spectre: Evolutionary Biology in the Modern World.* Princeton, N.J.: Princeton University Press.

Rosser, Sue V. 1992. *Biology and Feminism: A Dynamic Interaction.* New York: Twayne.

Russell, Bertrand. 1912. "The Philosophy of Bergson." *The Monist* 22.3: 321–347.

Russell, Bertrand. *The Philosophy of Bergson*. Cambridge: Bowes and Bowes.

——. 1957. *Mysticism and Logic.* Garden City, N.Y.: Doubleday.

Saussure, Ferdinand de. 1974. *Course in General Linguistics.* Trans. Charles Bally and Albert Secherhaye. New York: Fontana/Collins.

Sayers, Janet. 1982. *Biological Politics: Feminist and Anti-Feminist Perspectives.* London: Tavistock.

Schilder, Paul. 1978. *The Image and Appearance of the Human Body: Studies in the Constructive Energies of the Psyche.* New York: International Universities Press.

Schnelle, Helmut. 1994. "Turing Naturalized: Von Neumann's Unfinished Project." In *The Universal Turing Machine: A Half Century Survey*, ed. Rolf Herken, 539–559. Vienna: Springer-Verlag.

Shields, Rob. 2003. *The Virtual.* London: Routledge.

Simondon, Gilbert. 1993. "The Genesis of the Individual." In *Incorporations*, ed. J. Crary and S. Kwinter, 297–319. New York: Zone.

Smith, Adam. 1776. *An Inquiry into the Nature and Causes of the Wealth of Nations.* London: Condell.

——. 1982. *The Wealth of Nations.* Books 1–3. Harmondsworth, England: Penguin.

Spafford, Eugene H. 1992. "Computer Viruses: A Form of Artificial Life?" In *Artificial Life II*, ed. C. Langton, C. Taylor, J. D. Farmer, and S. Ramussen. Redwood City, Calif.: Addison-Wesley.

Spencer, Herbert. 1864. *The Principles of Biology.* London: Williams and Norgate.

——. 1872. *First Principles of a New System of Philosophy.* New York: Appelton.

Sterelny, Kim. 2001. *Dawkins v. Gould: Survival of the Fittest.* Oxford: Icon Books.

Stewart, J. McKellar. 1911. *A Critical Exposition of Bergson's Philosophy.* London: Macmillan.

Strong, Tracy B. 1988. *Friedrich Nietzsche and the Politics of Transfiguration.* Berkeley: University of California Press.

Swenson, Rod. 1997. "Thermodynamics, Evolution, and Behavior." In *The Encyclope-*

dia of Comparative Psychology, ed. G. Greenberg and M. Haraway. New York: Garland.

Thompson, D'Arcy W. 1959. *On Growth and Form*. Cambridge, England: Cambridge University Press.

Turetsky, Philip. 1998. *Time*. London: Routledge.

Turing, Alan. 1950. "Computing Machinery and Intelligence." *Mind* 59: 433–460.

——. 1952. "The Chemical Basis of Morphogenesis." *Philosophical Transactions of the Royal Society of London* B237: 37–72.

Volk, Tyler. 1995. *Metapatterns: Across Space, Time and Mind*. New York: Columbia University Press.

Waddington, Charles H. 1957. *The Strategy of the Genes: A Discussion of Some Aspects of Theoretical Biology*. London: George Allen and Unwin.

——. 1977. *Tools for Thought*. London: Jonathan Cape.

Waldrop, M. Mitchell. 1992. *The Emerging Science at the Edge of Chaos*. New York: Touchstone.

Wallace, Alfred Russel. 1870. *Contributions to the Theory of Natural Selection*. London: Macmillan.

——. 1889. *Darwinism: An Exposition of the Theory of Natural Selection*. London: Macmillan.

——. 1910. *The World of Life: A Manifestation of Creative Power, Directive Mind and Ultimate Purpose*. London: Chapman and Hall.

Wallace, Jeff. 1995. "Introduction: Difficulty and Defamiliarisation—Language and Process in *The Origin of Species*." In *Charles Darwin's The Origin of Species: New Interdisciplinary Essays*, ed. David Amigone and Jeff Wallace. Manchester, England: Manchester University Press.

Weismann, August. 1893. *The Theory of Heredity*. Trans. W. Newton Parker and H. Ronnfeldt. London: Walter Scott.

Whitford, Margaret. 1991. *Luce Irigaray: Philosophy in the Feminine*. London: Routledge.

Wilden, Anthony. 1972. *System and Structure: Essays in Communication and Exchange*. London: Tavistock.

Williams, George. 1966. *Adaptation and Natural Selection*. Princeton, N.J.: Princeton University Press.

Williams, Linda. 2001. *Nietzsche's Mirror: The World as Will to Power*. New York: Roman and Littlefield.

Wilson, Edward O. 1980. *Sociobiology: The Abridged Edition*. Cambridge, Mass.: Belknap Press of Harvard University Press.

Wolfram, Stephen. 2002. *A New Kind of Science*. Champaign, Ill.: Wolfram Media.

Wood, David. 2001. *The Deconstruction of Time*. Evanston, Ill.: Northwestern University Press.

INDEX

Elizabeth Grosz is a professor of women's and gender studies at Rutgers University. She is the author of *Architecture from the Outside: Essays on Virtual and Real Space* (2001) and editor of *Becomings: Essays in Time, Memory, Futures* (1999).

Library of Congress Cataloging-in-Publication Data
Grosz, E. A. (Elizabeth A.)
The nick of time : politics, evolution, and the untimely / Elizabeth Grosz.
p. cm.
Includes bibliographical references and index.
ISBN 0-8223-3400-3 (alk. paper)
ISBN 0-8223-3397-X (pbk. : alk. paper)
1. Time. 2. Body, Human. 3. Darwin, Charles, 1809–1882. 4. Nietzsche, Friedrich Wilhelm, 1844–1900. 5. Bergson, Henri, 1859–1941. I. Title.
BD638.G74 2004 128–dc22 2004011106